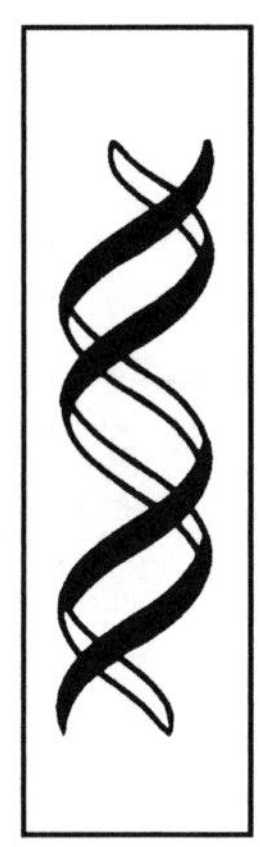

Volume 78

Advances in Genetics

Advances in Genetics, Volume 78

Serial Editors

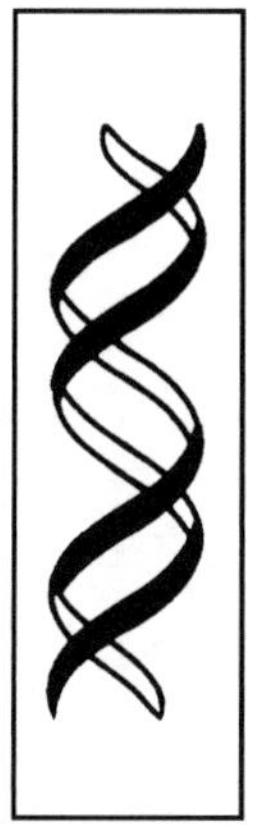

Volume 78

Advances in Genetics

Edited by

Stephen F. Goodwin

University of Oxford
Oxford, UK

Theodore Friedmann

University of California at San Diego
School of Medicine, USA

Jay C. Dunlap

The Geisel School of Medicine at Dartmouth
Hanover, NH, USA

AMSTERDAM • BOSTON • HEIDELBERG • LONDON
NEW YORK • OXFORD • PARIS • SAN DIEGO
SAN FRANCISCO • SINGAPORE • SYDNEY • TOKYO
Academic Press is an imprint of Elsevier

Academic Press is an imprint of Elsevier

525 B Street, Suite 1900, San Diego, CA 92101-4495, USA
225 Wyman Street, Waltham, MA 02451, USA
32 Jamestown Road, London, NW1 7BY, UK
Radarweg 29, POBox 211, 1000 AE Amsterdam, The Netherlands

First edition 2012

ISBN: 978-0-12-394394-1
ISSN: 0065-2660

Printed and bound in USA

12 13 14 11 10 9 8 7 6 5 4 3 2 1

Contents

Contributors

Numbers in parentheses indicate the pages on which the authors' contributions begin.

David J. Hosken (169) Centre for Ecology and Conservation, University of Exeter, Cornwall Campus, Tremough, Penryn, Cornwall, United Kingdom

Preetmoninder Lidder (1) Office of Knowledge Exchange, Research and Extension, Research and Extension Branch, Food and Agriculture Organization of the UN (FAO), Viale delle Terme di Caracalla, Rome, Italy

Wayne G. Rostant (169) Centre for Ecology and Conservation, University of Exeter, Cornwall Campus, Tremough, Penryn, Cornwall, United Kingdom

Andrea Sonnino (1) Office of Knowledge Exchange, Research and Extension, Research and Extension Branch, Food and Agriculture Organization of the UN (FAO), Viale delle Terme di Caracalla, Rome, Italy

Nina Wedell (169) Centre for Ecology and Conservation, University of Exeter, Cornwall Campus, Tremough, Penryn, Cornwall, United Kingdom

Contributors

1 Biotechnologies for the Management of Genetic Resources for Food and Agriculture

Preetmoninder Lidder and Andrea Sonnino

Office of Knowledge Exchange, Research and Extension, Research and Extension Branch, Food and Agriculture Organization of the UN (FAO), Viale delle Terme di Caracalla, Rome, Italy

Advances in Genetics*, *Vol. 78 0065-2660/12 $35.00

Published by Elsevier Inc. http://dx.doi.org/10.1016/B978-0-12-394394-1.00001-8

B. "Omics"
C. Bioinformatics
D. Cryopreservation
E. *In vitro* slow growth storage
F. Micropropagation
G. Chromosome set manipulation
H. Molecular marker-assisted selection
I. Mutagenesis
J. Transgenesis

V. Current Status of Biotechnologies for the Management of Animal Genetic Resources
A. Molecular markers
B. "Omics"
C. Bioinformatics
D. Cryopreservation
E. Reproductive biotechnologies
F. Biotechnologies for disease diagnosis and prevention
G. Molecular marker-assisted selection
H. Mutagenesis
I. Transgenesis

VI. Current Status of Biotechnologies for the Management of Aquatic Genetic Resources
A. Molecular markers
B. "Omics"
C. Bioinformatics
D. Cryopreservation
E. Reproductive biotechnologies
F. Chromosome set manipulation
G. Biotechnologies for disease diagnosis and prevention
H. Molecular marker-assisted selection
I. Transgenesis

VII. Current Status of Biotechnologies for the Management of Microbial Genetic Resources
A. Molecular markers
B. "Omics"
C. Bioinformatics
D. Cryopreservation
E. Pathogen detection in food
F. Food preservation and production of food and feed ingredients
G. Biofertilizers

ABSTRACT

In recent years, the land area under agriculture has declined as also has the rate of growth in agricultural productivity while the demand for food continues to escalate. The world population now stands at 7 billion and is expected to reach 9 billion in 2045. A broad range of agricultural genetic diversity needs to be available and utilized in order to feed this growing population. Climate change is an added threat to biodiversity that will significantly impact genetic resources for food and agriculture (GRFA) and food production. There is no simple, all-encompassing solution to the challenges of increasing productivity while conserving genetic diversity. Sustainable management of GRFA requires a multipronged approach, and as outlined in the paper, biotechnologies can provide powerful tools for the management of GRFA. These tools vary in complexity from those that are relatively simple to those that are more sophisticated. Further, advances in biotechnologies are occurring at a rapid pace and provide novel opportunities for more effective and efficient management of GRFA.

Biotechnology applications must be integrated with ongoing conventional breeding and development programs in order to succeed. Additionally, the generation, adaptation, and adoption of biotechnologies require a consistent level of financial and human resources and appropriate policies need to be in place. These issues were also recognized by Member States at the FAO international technical conference on Agricultural Biotechnologies for Developing Countries (ABDC-10), which took place in March 2010 in Mexico. At the end of the conference, the Member States reached a number of key conclusions, agreeing, *inter alia*, that developing countries should significantly increase sustained investments in capacity building and the development and use of biotechnologies to maintain the natural resource base; that effective and enabling national biotechnology policies and science-based regulatory frameworks can facilitate the development and appropriate use of biotechnologies in developing countries; and that FAO and other relevant international organizations and donors should significantly increase their efforts to support the strengthening of national capacities in the development and appropriate use of pro-poor agricultural biotechnologies.

ABBREVIATIONS

ABDC-10	FAO international technical conference on agricultural biotechnologies in developing countries
AFLP	amplified fragment length polymorphism
AI	artificial insemination
AMF	arbuscular mycorrhizal fungi
AW-IPM	area-wide integrated pest management
BWB	breeding without breeding
CBD	Convention on Biological Diversity
CBOL	Consortium for the Barcode of Life
CBPP	contagious bovine pleuropneumonia
cDNA	complementary DNA
C-ELISA	competitive ELISA
CGIAR	Consultative Group on International Agricultural Research
CIAT	International Centre for Tropical Agriculture
CIP	International Potato Center
CSV	classical swine fever
CWR	crop wild relatives
DArT	diversity arrays technology
DGGE	denaturing gradient gel electrophoresis
DH	doubled haploid
DIVA vaccine	vaccine capable of differentiating infected and vaccinated animals
ELISA	enzyme-linked immunosorbent assay
EST	expressed sequence tag
ET	embryo transfer
EU	European Union
FAO	Food and Agriculture Organization of the United Nations
FMD	foot and mouth disease
GCP	Generation Challenge Program
GMO	genetically modified organism
GnRH	gonadotropin-releasing hormone
GRFA	genetic resources for food and agriculture

GS	genomic selection
IAEA	International Atomic Energy Agency
IHNV	infectious hematopoietic necrosis virus
IITA	International Institute of Tropical Agriculture
INIBAP	International Network for the Improvement of Banana and Plantain
IPM	integrated pest management
IPR	intellectual property rights
ISAG	International Society of Animal Genetics
ISPM	International Standards for Phytosanitary Measure
IVF	*in vitro* fertilization
KHV	koi herpesvirus
LAMP PCR	loop-mediated isothermal amplification PCR
MAS	marker-assisted selection
NACA	Network of Aquaculture Centres in Asia-Pacific
N_e	effective population size
NERICA	New Rice for Africa
NGS	next generation sequencing
OIE	World Organization for Animal Health
PCR	polymerase chain reaction
PPR	peste des petits ruminants
PSB	P-solubilizing bacteria
PSF	P-solubilizing fungi
QPM	quality protein maize
QTL	quantitative trait locus
RAPD	random amplified polymorphic DNA
RFLP	restriction fragment length polymorphism
RIA	radioimmunoassay
RP	rinderpest
rRNA	ribosomal RNA
RT-PCR	reverse transcriptase PCR
SE	somatic embryogenesis
SIT	sterile insect technique
SNP	single nucleotide polymorphism
TAAPs/TAADs	transboundary aquatic animal pathogens/diseases
TADs	transboundary animal diseases
TGGE	temperature gradient gel electrophoresis
TILLING	targeting induced local lesions IN genomes
T-RFLP	terminal-RFLP
UN	United Nations
WSSV	white spot syndrome virus

I. INTRODUCTION

The convention on biological diversity (CBD) defines genetic resources as *genetic material of actual or potential value*. Genetic resources for food and agriculture (GRFA) are the raw material for agricultural development and therefore their sustainable use is crucial for global food security, particularly for the rural poor in developing countries. Furthermore, genetic resources are not uniformly distributed on the planet, being richer in the tropical and subtropical zones. Consequently, countries and regions are interdependent and benefit sharing and international cooperation are essential.

The world's food and agriculture production depends upon plant, animal, aquatic, forest, microbial, and invertebrate genetic resources. Although many plants are edible and over 7000 species have been cultivated or collected for food, only 30 crops provide 95% of human dietary energy needs with just three of them, that is, rice, wheat, and maize, providing more than 50% (FAO, 1997). It should be noted that although the genetic diversity within this small number of major crop species is quite substantial, most crops exhibit less genetic variation than their wild relatives due to domestication bottlenecks (Tanksley and McCouch, 1997). For livestock, of the 50,000 known avian and mammalian species, about 40 have been domesticated, with fewer than 14 species accounting for over 90% of global livestock production (FAO, 1999, 2007a). Within these species, a great variety of breeds have been developed since domestication, and genetic diversity in livestock breeds is usually much greater than that in crop varieties. However, 21% of the 8054 breeds reported globally are at risk of extinction.[1]

For aquatic and forest genetic resources, wild populations play an important role in addition to domesticated populations. More than 300 aquatic species (excluding aquatic plants) are farmed worldwide, while over 1000 species are harvested from capture fisheries.[2] However, aquaculture is currently the fastest growing animal food-producing sector and improved or domesticated strains are becoming increasingly important (FAO, 2010a). The majority of the world's forest genetic resources are unknown, found in largely unmanaged and undomesticated forests, with only 7% being in plantations (FAO, 2010b). It is estimated that the number of tree species varies from 80000 to 100000, of which fewer than 500 have been studied in detail.[3] Finally, microorganisms (bacteria, fungi, and viruses) and invertebrates (insects, spiders, and earthworms) include numerous species with great genetic diversity that are invaluable contributors to agro-ecosystems.[4]

[1] http://www.fao.org/docrep/meeting/021/am131e.pdf.

[2] http://www.fao.org/fishery/statistics/software/fishstat/en.

[3] http://www.fao.org/nr/cgrfa/cthemes/forest/en/.

[4] ftp://ftp.fao.org/docrep/fao/meeting/017/ak534e.pdf.

Genetic resources are vulnerable to losses and are rapidly dwindling due to habitat change/degradation, overexploitation, pollution, invasive alien species, and climate change. According to the 2011 update of the International Union for Conservation of Nature Red List of Threatened Species™, 19,265 of the 59,508 assessed species are threatened with extinction.[5] The UN General Assembly declared 2010 as the International Year of Biodiversity[6] to coincide with the 2010 Biodiversity Target[7] of achieving a significant reduction in the rate of biodiversity loss. However, this target has not been met. These shortcomings apply equally to agricultural and wild biodiversity. In fact, as confirmed in the third edition of the Global Biodiversity Outlook,[8] genetic diversity continues to decline in agricultural systems.

As outlined above, the world's food supply is dependent on a small number of species and the continuing reduction in genetic diversity is a major cause for concern. Maintaining genetic diversity is essential since removal of a single species can affect the functioning of global ecosystems. Further, genetically diverse populations have a greater adaptive potential and are more resilient to environmental changes, in addition to being a source of economically and scientifically important traits. Thus conservation of genetic diversity is crucial to ensure its continued availability for adaptation to climate change and future production, market, and societal needs. Conservation is also important as it provides insurance against unforeseen catastrophic events and preserves genetic resources of cultural and historical value.

Since GRFA are finite and once lost cannot be regained, proper management is fundamental. Biotechnology applications can provide comparative advantages over, or can increase the effectiveness of, traditional technologies for the characterization, conservation, and utilization of GRFA. Indeed, both the *Global Plan of Action for the Conservation and Sustainable Utilization of Plant Genetic Resources for Food and Agriculture*[9] and the *Global Plan of Action for Animal Genetic Resources*[10] identify a direct or indirect role for agricultural biotechnologies in some of their priority areas.

In this document, the following definition of biotechnology, based on that of the CBD, is used, that is, "any technological application that uses biological systems, living organisms, or derivatives thereof, to make or modify products or processes for specific use".[11] Some of the biotechnologies described, for example, the use of molecular markers may be applied to all the agricultural

[5]http://www.iucnredlist.org/documents/summarystatistics/2011_1_RL_Stats_Table_1.pdf.
[6]http://www.cbd.int/2010/.
[7]http://www.cbd.int/2010-target/.
[8]http://www.cbd.int/doc/publications/gbo/gbo3-final-en.pdf.
[9]ftp://ftp.fao.org/docrep/fao/meeting/015/aj631e.pdf.
[10]ftp://ftp.fao.org/docrep/fao/010/a1404e/a1404e00.pdf.
[11]http://www.cbd.int/convention/articles.shtml?a=cbd-02.

sectors, that is, they are cross-sectoral, while others are more sector specific, such as tissue culture (in crops and forest trees), embryo transfer (ET) (livestock), and sex-reversal (fish). Issues such as the health or environmental risks of any biotechnology product are beyond the scope of this chapter.

The chapter aims to provide an updated review of the current state of research, development, and application of the many biotechnologies that are currently available and that can be employed for the management of GRFA. It is intended to provide information and insights that can both assist countries to establish, maintain, and advance policies in relation to biotechnologies in this area and help to identify gaps which might merit further studies. Use of biotechnologies, particularly advanced biotechnologies, requires a commitment of financial and human resources. However, detailed discussions on the relative merits of applying the different biotechnologies, compared with each other or with conventional technologies, are not entered into here, as they are often case specific and are influenced by variable factors such as the current costs and technical feasibility of the different technologies.

This document is divided into six main sections. Section II presents an overview of biotechnology applications relevant for the characterization, conservation, and utilization of GRFA, followed by sector-specific sections (Sections III–VII) that provide details pertinent for the management of crop, forest, animal (livestock), aquatic (fisheries and aquaculture), and microbial GRFA, respectively. Section VIII provides conclusions and outlook.

II. BIOTECHNOLOGIES APPLIED TO THE CHARACTERIZATION, CONSERVATION, AND UTILIZATION OF GENETIC RESOURCES FOR FOOD AND AGRICULTURE

For biotechnology applications to succeed, they should complement conventional conservation and breeding activities and build upon existing and active programs. Conventional breeding has provided enormous benefits in the past and will continue to do so in the future. For example, domestication of aquatic and forest species is relatively recent and they have not been genetically improved to the same extent as crop and livestock species. They may, therefore, derive particular benefits from the use of conventional breeding. Further, relevant components of production and market systems, as well as socioeconomic, environmental, and cultural considerations, should also be taken into account before the decision to apply a particular biotechnology is made. Biotechnologies *per se* are not the solution, but when integrated with ongoing, appropriately designed conservation, breeding, and development programs, can be of significant assistance in meeting the needs of an expanding and increasingly urbanized population while maintaining the diversity of genetic resources.

A. Characterization of GRFA

Characterization is a prerequisite for identifying and prioritizing the genetic resources to be conserved and is fundamental for optimizing appropriate allocation to conservation programs when funding resources are limited (Boettcher *et al.*, 2010). Characterization also links conservation and utilization as it allows the identification of unique and valuable traits of conserved genetic resources, both *in situ* and *ex situ*, for incorporation into breeding programs. In addition, characterization of GRFA is essential for ensuring ownership of GRFA (to promote and control bioprospecting, that is, prospecting for commercially valuable biological or genetic resources and the accompanying traditional knowledge, and to avoid biopiracy, that is, the illegal appropriation of these resources) as well as access to and fair and equitable sharing of their benefits, especially in conjunction with intellectual property rights (IPR) management.

Genetic resources can be characterized with respect to genotypes, phenotypes, morphological traits, measures of genetic diversity, genetic distance, population size and structure, geographical distribution, and degree of endangerment. Biotechnology applications for characterization include molecular markers and the so-called "omic"[12] technologies.

1. Molecular markers (cross-sectoral)

Molecular markers are heritable, identifiable DNA sequences that are found at specific locations within the genome and can be used to detect DNA polymorphism. The first widely used markers were isozymes[13] and they are still being applied today (e.g., to characterize forest trees). However, isozymes show low levels of polymorphism and relatively low abundance and, in many instances, have been replaced by more sensitive techniques.

The assay of molecular markers requires only small amounts of biological material that can be easily transported and stored. Since they are fixed at fertilization, molecular markers are not affected by environmental conditions. They can be used at any growth stage, which is especially advantageous for long-lived species such as forest trees. Additionally, molecular markers can be selectively neutral or associated with functional variation. Neutral markers are particularly useful for understanding population size and structure, estimating relationships between populations and hybrid identification, while functional markers allow specific alleles within pedigrees and populations to be tracked.

[12]"Omics" is a general term for a broad discipline of science and engineering for analyzing the interactions of biological information objects in various "omes," such as the genome, proteome, etc.

[13]An isozyme is a genetic variant of an enzyme. Isozymes for a given enzyme share the same function but may differ in level of activity, as a result of minor differences in their amino acid sequence (FAO, 2001).

Different kinds of molecular marker systems are available such as restriction fragment length polymorphisms (RFLPs), random amplified polymorphic DNAs (RAPDs), amplified fragment length polymorphisms (AFLPs), microsatellites, and single nucleotide polymorphisms (SNPs) (FAO, 2001). RFLPs are based on the detection of variation in the length of fragments generated when DNA is treated with restriction endonucleases. RAPDs are detected using the polymerase chain reaction (PCR), that is, the genomic template is amplified with single, short (usually 10-mer) randomly chosen primers. AFLPs are generated by the PCR amplification of restriction endonuclease-treated DNA ligated to an adaptor sequence. Microsatellites are segments of DNA, around five or fewer bases, that are repeated a variable number of times in tandem and are detected by PCR. SNPs are genetic markers resulting from single base changes at particular positions within a DNA sequence.

The aforementioned markers differ with respect to technical requirements, the amount of time, money, and labor needed, reproducibility, level of polymorphism detected, type of expression (dominant or codominant), and the number of genetic markers that can be detected throughout the genome (reviewed in Budak *et al.*, 2004; Duran *et al.*, 2009; FAO, 2003, 2007b; Okumus and Ciftci, 2003; Schlötterer, 2004; Spooner *et al.*, 2005; Teneva, 2009). No single molecular marker system is suitable for all kinds of studies, and choice of the marker system used should be based on its appropriateness for the information required, the population/species being studied, and the available financial and operational resources. For example, SNPs and microsatellites are excellent choices for parentage analysis while mitochondrial DNA and Y chromosome markers are useful for the identification of maternal and paternal lineages, respectively. Molecular marker information should also be used in conjunction with other information sources (e.g., phenotypic traits, population data, and evaluation of the production system) to assist decision-making regarding conservation, especially since populations with low genetic diversity at the molecular level may have unique phenotypes valuable for conservation (Peter *et al.*, 2007).

Irrespective of the kind of molecular marker used, technical infrastructure and know-how, as well as relatively expensive consumables, are necessary, although multiplexing for PCR-based markers can significantly improve the speed and efficiency of genotyping leading to reduced costs and labor. However, since marker development costs are higher than running (that is, typing using known markers) costs, research and development, including in developing countries, may benefit from the large number of markers already available for many species.

Molecular markers can be used to characterize GRFA in a variety of different ways, namely, for the

(a) Assessment of intraspecific genetic diversity: Molecular markers provide valuable information about intraspecific genetic diversity, that is, within and between populations of the same species, in all the agricultural sectors. Biological tissue (e.g., blood, leaves, fish fin clips) is sampled from a representative set of individuals from a number of populations of interest, typed for marker loci followed by a statistical analysis of the data to estimate the genetic distance between and genetic variability within the populations being studied (Ruane and Sonnino, 2006). Determination of the amount of genetic diversity present within or between populations as well as the structure and geographic distribution of diversity in populations can aid in the identification of the most suitable strategies for conservation and utilization.

(b) Documentation of domestication events: Assessment of genetic distances has contributed to the understanding of the history and timing of domestication by identifying wild populations most closely related to domesticated species as well as the putative centres of origin.

(c) Detection of interspecific variation: When species are difficult to identify morphologically, detection of interspecific variation is particularly significant, for example, for the forestry and fisheries sectors. DNA barcoding is an emerging tool for cataloging biodiversity that uses a standardized gene region for species identification and discovery. It is being applied on a global scale through an international initiative, the International Barcode of Life project[14] that aims to barcode 5 million specimens representing 500,000 species. Additional community-based initiatives include the Consortium for the Barcode of Life (CBOL),[15] that promotes barcoding through information exchange, conferences, outreach, and training, and the European Consortium for the Barcode of Life.[16] Two central DNA barcode databases exist, the Barcode of Life Data Systems,[17] which aids collection, management, analysis, and use of DNA barcodes, and the International Nucleotide Sequence Database Collaboration.[18]

(d) Estimation of effective population size (N_e): The effective population size (N_e) is a key parameter in the field of population genetics and is defined as the number of individuals in an idealized population that would give rise to the same inbreeding rate or the same amount of genetic drift as in the real population of interest. N_e is generally smaller (often much smaller) than the actual population size and is an important indicator for determining the degree of endangerment of a population. N_e is affected by several factors

[14]http://www.barcodeoflife.org/content/about/what-ibol.
[15]http://www.barcodeoflife.org/content/about/what-cbol.
[16]http://www.ecbol.org/.
[17]http://www.boldsystems.org/views/login.php.
[18]http://www.insdc.org/.

such as sex ratio, numbers of offspring for individual matings, variable population size, mating systems, and selection (Caballero, 1994). N_e is traditionally estimated using pedigree data, censuses, and genotype data (Bartley *et al.*, 1992), but this information may be difficult to obtain for wild populations (Ruane and Sonnino, 2006).

(e) Investigation of gene flow between domesticated populations and their wild relatives: Gene flow from domesticated populations can influence the genetic variability of fitness traits in wild populations and in some cases lead to outbreeding depression [loss of genotypes important for local adaptation and reduction in reproductive fitness in the first or later generations following attempted crossing of populations (Frankham *et al.*, 2011)]. Molecular markers can be used as tools to conserve wild relatives by distinguishing hybrids from nonhybrids, allowing for selection against the hybrids and purging of their genes from the wild population. Markers can also be utilized to preserve cultivated species by detecting introgression of genes from wild relatives and loss of domestication-related traits.

(f) Identification of quantitative trait loci (QTLs)[19]: Molecular markers are routinely used for tracing and mapping alleles of interest in a segregating population. Most economically important traits (e.g., yield, disease resistance) are quantitative, that is, typically controlled by many genes that have an additive effect and also strongly influenced by the environment. Molecular markers are highly effective tools for the identification of these QTLs. Markers can be employed to construct genetic linkage maps,[20] comprising markers interspersed at short, regular intervals throughout the genome, with the distances between markers reflecting the degree of linkage. The putative QTLs are then mapped to a small region on the chromosome followed, in some cases, by the identification of the genes responsible for the trait using a positional cloning[21]/candidate gene[22] approach. It should be noted, however, that for QTL identification a major challenge is the technical sophistication associated with creating mapping populations, recording meaningful phenotypes, and compiling genetic maps.

[19] A quantitative trait locus is a locus where allelic variation is associated with variation in a quantitative trait, such as yield, tolerance to abiotic stresses, etc. (FAO, 2001).

[20] A linkage map is a linear or circular diagram that shows the relative positions of genes on a chromosome as determined by recombination fraction (FAO, 2001).

[21] A strategy for gene cloning that relies on the identification of closely linked markers to the target trait and then uses chromosome walking to identify, isolate, and characterize the gene(s) responsible for the trait (FAO, 2001).

[22] A gene known to be located in the same region as a DNA marker that has been shown to be linked to a single-locus trait or to a QTL, and whose deduced function suggests that it could be the source of genetic variation in the trait in question (FAO, 2001).

(g) Conservation management of genetic resources: Molecular markers are of considerable value in the conservation of plant genetic resources in defining conservation strategies, gap analysis studies, and developing sampling strategies for gene banks for prioritizing populations for conservation. Molecular markers can also be used for the management of conserved germplasm by increasing the efficiency of gene bank operations, especially in characterization and regeneration activities.

2. "Omic" technologies (cross-sectoral)

Genomics refers to the study of an organism's genome at the DNA level. To date, the genomes of more than 1000 organisms, including plants, animals, fish,[23] forest trees, microorganisms, and invertebrates, have been sequenced. Outputs from genome sequencing can be further enhanced by elucidating patterns of gene expression and gene function through functional genomic technologies such as transcriptomics, proteomics, and metabolomics, that is, mRNA, protein, and metabolite profiling, respectively, thus providing a thorough gene inventory (Fears, 2007; Schneider and Orchard, 2011). Additionally, proteomic and metabolomics are not reliant on having a preavailable genome sequence. "Omic" information, analyzed in conjunction with bioinformatics, can be exploited to characterize and utilize GRFA in novel ways, since entire networks of genes (as opposed to single genes) can be analyzed spatially and/or temporally and in a much speedier manner compared to conventional technologies.

Genomic information has the potential to contribute to and solve some key problems in the management of genetic resources (reviewed in Allendorf *et al.*, 2010). For example, the accelerated generation of molecular markers throughout the genome can increase the precision and accuracy for estimating and monitoring N_e. The creation of high-density linkage maps can greatly assist QTL mapping and cloning of the corresponding genes due to improved reliability of data and conclusions, making selection for quantitatively inherited characters quicker and more precise.

Specialized fields of genomics are integrating information from numerous sources to benefit from the immense amount of genomic data available and to maximize the characterization benefits of genomic approaches. Comparative genomics takes advantage of synteny[24] and the conserved functions of genes, regulatory and noncoding sequences, and has facilitated the prediction

[23]The term also includes aquatic invertebrates, for example, molluscs, echinoderms, and crustaceans.

[24]The occurrence of two or more loci on the same chromosome, without regard to their genetic linkage. The term is increasingly used to describe the conservation of gene order between related species (FAO, 2001).

of candidate genes in close relatives following the availability of the sequences of model/key species (especially important for utilizing the genetic diversity of under-resourced and orphan crop species and less-common livestock species).

Population genomics utilizes genome-wide patterns of sequence variation, at the population level, to detect genes subjected to strong selection pressure, thus contributing to the understanding of adaptive evolution (FAO, 2010c). Genomic scans for detection of "selection signatures" (that is, increased frequency of nucleotides linked to favourable mutations) can provide insight into the evolutionary history of species and their subdivision into distinct breeds and varieties, with comparisons between species being more informative for the identification of older events, and between and within-populations for revealing more recent episodes of selection.

Landscape genomics combines genomic information with geoenvironmental data to determine, for example, which breeds of livestock are best suited for certain production circumstances (FAO, 2007a). Xenogenomics, that is, functional genomics specifically targeting nonmodel and noncrop plants with enhanced tolerance to abiotic stresses, is being used to discover novel genes from indigenous and exotic plant species living in extreme conditions (John and Spangenberg, 2005). Metagenomics, the study of the genomes of samples taken directly from the environment, for instance, soil samples, is a relatively new field of genetic research that enables studies of unculturable organisms as well as characterization of biodiversity at the ecosystem level (Marco, 2010).

Finally, genomics can be utilized as a tool for bioprospecting indigenous GRFA for economically important traits. Such knowledge is key for developing countries not only to capitalize on the full potential of their genetic resources for economic and conservation benefits but also to prevent the exploitation of their rich biodiversity and avoid unfair sharing of benefits. High-throughput genomic approaches have been utilized by a few developed countries and the private sector, especially for microbial bioprospecting in the Antarctic and Arctic ecosystems (Leary and Walton, 2010; Peck *et al.*, 2005).

High levels of financial investments and expertise are required for the setup and maintenance of laboratories/centers capable of providing "omic" facilities. Further, trained scientists, good internet access, and computer facilities are vital to take advantage of the publicly available sequence information and bioinformatic tools. Consequently, these technologies are being used only in some cases across the different agricultural sectors in developing countries. Nevertheless, genomic sequencing is becoming steadily cheaper. High-throughput SNP genotyping costs have decreased by up to 10-fold while data throughput has increased by the same magnitude (Ribaut *et al.*, 2010).

3. Bioinformatics (cross-sectoral)

The availability of enormous amounts of data generated by the high-throughput "omic" technologies has necessitated the development of methods for processing, analyzing, integrating, and interpreting the data. Bioinformatics refers to the research, development, and application of computational and statistical tools and information processing methods for the management of biological information. Since bioinformatics is interdisciplinary in nature and blends many fields together, collaboration and synergy between biologists, biostatisticians, and bioinformaticists is key for successful data mining, that is, the extraction of relevant information from large datasets.

Bioinformatic tools can be used for molecular marker discovery/ prediction, sequence analyses including prediction of function, genome and chromosome annotation, phylogeny estimation, data mining for genes of interest, high-throughput analysis of gene expression, systems biology, and population genetics. Numerous specialized public databases exist. The three major sequence databases are the National Centre for Biotechnology Information,[25] the European Molecular Biology Laboratory,[26] and the DNA Databank of Japan,[27] which are also the primary sites that provide access to a collection of bioinformatic tools. Microarray and gene expression data can be accessed at the Gene Expression Omnibus,[28] the ArrayExpress,[29] and the Center for Information Biology Gene Expression Database.[30] AgBase[31] provides functional modeling resources for agriculturally important animal, plant, microbe, and parasite genomes (McCarthy *et al.*, 2011).

At the protein level, annotated protein sequences are provided by UniProt[32] and bio-macromolecular structure data can be accessed in the Protein Data Bank,[33] protein modifications in UniMod[34] and RESID,[35] and protein interactions in the International Molecular Exchange Consortium databases.[36]

[25]http://www.ncbi.nlm.nih.gov/Tools/.
[26]http://www.ebi.ac.uk/Tools/.
[27]http://www.ddbj.nig.ac.jp.
[28]http://www.ncbi.nlm.nih.gov/geo/.
[29]http://www.ebi.ac.uk/arrayexpress/.
[30]http://cibex.nig.ac.jp.
[31]http://www.agbase.msstate.edu/.
[32]http://www.uniprot.org.
[33]http://www.wwpdb.org/.
[34]http://www.unimod.org/.
[35]http://www.ncifcrf.gov/resid/.
[36]http://www.imexconsortium.org/.

Proteomics data repositories (reviewed in Vizcaíno *et al.*, 2010) include the Global Proteome Machine Database,[37] PeptideAtlas,[38] the PRoteomics IDEntifications database,[39] and Tranche.[40]

Currently, the ability to generate genomic information and the trend at which this capacity is increasing threatens to outpace the corresponding progression in development of hardware and tools for data management, storage, and analysis. Bioinformatics may thus be a limiting factor in the full exploitation of genomics for the management of GRFA, and this may be particularly true in developing countries.

B. Conservation of GRFA

Two major strategies exist for conservation.[41] *In situ* conservation allows continued evolution and adaptation of a species in response to the environment. Albeit more dynamic, it is exposed to habitat destruction by natural calamities and/or human interference. *Ex situ* conservation can be used to ensure easy and ready accessibility of reproductive material. Gene banks constitute the most significant and widespread means of conserving plant genetic resources. Currently, 1750 gene banks worldwide maintain 7.4 million accessions, with the large majority of the accessions in the form of seeds and the largest collections held by the CGIAR centers in the public domain (FAO, 2010c). The methods described in the preceding section are of utility in monitoring conserved species and/or populations, both *in situ* and *ex situ*.

An effective link between both strategies is important. For example, *ex situ* collections can be used for *in situ* population enhancement (by introducing genetic diversity and thus reducing inbreeding levels), or even to reintroduce rare/extinct species to the wild (Dulloo, 2011; Engels *et al.*, 2008). The two strategies are thus complementary rather than alternatives, and effective conservation strategies often incorporate elements of both, to devise the best strategy taking into account the biology of the species to be conserved, technical and financial aspects as well as infrastructural and human resources available (CBSG, 2011; Dulloo *et al.*, 2005; Volis and Blecher, 2010).

[37]http://gpmdb.thegpm.org/.

[38]http://www.peptideatlas.org/.

[39]http://www.ebi.ac.uk/pride/.

[40]http://www.tranche.proteomecommons.org.

[41]According to the CBD and the International Treaty on Plant Genetic Resources for Food and Agriculture (ITPGRFA), *in situ* conservation means the conservation of ecosystems and natural habitats and the maintenance and recovery of viable populations of species in their natural surroundings and, in the case of domesticated or cultivated species, in the surroundings where they have developed their distinctive properties. *Ex situ* conservation means the conservation of components of biological diversity outside their natural habitats.

1. Cryopreservation (cross-sectoral)

Cryopreservation involves the storage of germplasm at ultra-low temperatures (usually in liquid nitrogen at −196 °C), whereby all metabolic activities are suspended. It is a cost–effective option that allows for long-term storage, reduces the risk of loss (from diseases, disasters, etc.), requires limited space and minimal maintenance, and can offer extinct/selected genetic material for improved breeding in the future but does not allow for continued genetic adaptation. Its routine use is restricted in developing countries since the availability of economically priced liquid nitrogen is a particular constraint, although it can prove to be a more cost–effective method in the long term, provided the initial investment of a cryo facility is made (Dulloo *et al.*, 2009).

Cryopreservation is a useful method for long-term storage of germplasm, especially for plant species that are difficult to conserve as seeds due to low dessication tolerance. For aquatic species, cryopreservation has limited application because female gametes and fertilized eggs usually cannot be frozen. For livestock, the term "cryoconservation" is often used to refer to cryopreservation of germplasm for the purpose of genetic conservation, whereas cryopreservation refers to the actual freezing technology and its general application. This distinction is more relevant for livestock than for other sectors, because cryopreservation is more widely applied for uses other than conservation.

2. *In vitro* slow growth storage (crops and forest trees)

For crop and forest genetic resources, the majority of the accessions are maintained as seeds in gene banks. A significant number of crop and forest species do not produce orthodox seeds[42] and storage of their germplasm is, therefore, difficult. Other species, such as root and tuber crops, fruit trees, and forest trees, are vegetatively propagated because clonal multiplication allows the preservation of their unique genetic setups. In both cases, the germplasm can be preserved in field gene banks or *in vitro*.

Field gene banks are expensive to maintain, require more space, and are not very secure. Hence, short- to medium-term (1–15 years) conservation of vegetatively propagated crops and forest trees is best achieved with *in vitro* slow growth storage, that is, as sterile tissue/plantlets on nutrient gels. Growth is usually limited by reducing temperature (0–5 °C for cold-tolerant species and 15–20 °C for tropical species) and/or light intensity, by modifying the nutrients in the culture medium, reducing oxygen levels (Rao, 2004) and the use of mineral oil overlay (Mathur *et al.*, 1991).

[42]Seeds which can be dried to a low moisture content and stored at low temperatures without losing their viability over long periods of time.

Advantages of this method include reduced storage space for maintaining a large number of explants in an aseptic environment, decreased need for frequent subculturing, the potential for high clonal multiplication rates, ease of transfer of germplasm due to the smaller size of cultured material, and reduced need for quarantine during germplasm movement and exchange. However, *in vitro* maintenance is time and labor intensive, requires specialized equipment, and has an increased risk of somaclonal variation[43] as well as losses due to contamination of the culture media or mislabelling (Panis and Lambardi, 2006). A prerequisite, of course, is the availability of suitable protocols for *in vitro* culture in the species of interest. This is currently a major limitation for the application of the technology to a large number of forest tree species.

3. Reproductive biotechnologies (livestock and fisheries/aquaculture)

A number of reproductive biotechnologies have considerable potential for conserving livestock and fish by facilitating the storage, and eventual multiplication and dissemination, of genetic resources and reducing the risk of disease transmission. In livestock, however, the main use of these technologies is not for this purpose but as tools to increase animal production by increasing the number of offspring from elite individuals or by expanding the traditional geographical range of the most productive commercial breeds. This latter application can lead to the loss of indigenous breeds.

a. Controlled breeding (artificial breeding)

Controlled breeding is widely practised in aquatic species. Adults are spawned under controlled conditions, and often hormones or environmental stimuli are provided to promote spawning. Eggs or larvae are collected for placement alive in contained or open environments. Gametes may also be irradiated or chemically treated to destroy DNA of one sex (androgenesis and gynogenesis), thus allowing manipulation of the embryonic genome.

b. Artificial insemination

Artificial insemination (AI) is the process of collecting sperm cells from a donor male and manually depositing them into the reproductive tract of an ovulating female to achieve pregnancy. Progesterone monitoring and estrus synchronization are prerequisites for improving AI efficiency in livestock and ultimately the reproductive potential of animals.

[43]The term refers to epigenetic or genetic changes induced during the callus phase of plant cells cultured *in vitro*.

c. Progesterone monitoring

On their own, traditional methods for recording the reproductive cycle of the animal are often inaccurate. Combined with methods such as radioimmunoassay (RIA) and the enzyme-linked immunosorbent assay (ELISA), that are used to measure the progesterone level in blood or milk, a more precise assessment of the reproductive status of animals can be made to identify individuals that are anestrous or nonpregnant.

RIA is based on the use of a radioactively labeled antibody, where the amount of radiation detected indicates the amount of target substance present in the sample (FAO, 2001). ELISA is an antibody-based technique for the diagnosis of the presence and quantity of specific molecules in a mixed sample (FAO, 2001). An enzyme is attached to a reactant (antigen or antibody) in a system that is used to generate a color signal with an appropriate chromogen and substrate combination.

d. Estrus synchronization

Eestrus synchronization is the process of bringing females into estrous at the desired time by treatment with hormones such as estradiol, prostaglandins, progesterone, or gonadotropins. Such systems often use a combination of hormones to manipulate either the luteal or follicular phase of the estrus cycle while maintaining normal fertility.

e. Embryo transfer

ET involves the transfer of an embryo from a superior donor female to a less valuable female animal. Embryos for eventual transfer or cryopreservation are normally recovered using multiple ovulation (where the females are hormonally induced to release several oocytes). An alternative is to harvest oocytes from live donor animals (ovum pickup) or ovaries of slaughtered animals and then fertilize the oocytes *in vitro*, after which the fertilized embryos can be transferred fresh or frozen. On average, the gestation rates of frozen embryos are much lower than those of fresh embryos produced *in vitro*.

f. *In vitro* fertilization

In vitro fertilization (IVF) involves harvesting the oocytes from the donor's ovaries, culturing them, and fertilizing them with sperm *in vitro*, followed either by transfer to recipient females or cryopreservation. IVF is widely practiced in aquatic species where gametes are collected from both sexes and physically mixed to induce fertilization. Fertilized eggs or hatched larvae may be placed alive in contained or open environments.

g. Cloning

Cloning can be accomplished using embryo splitting, somatic cell nuclear transfer, or stem cell nuclear transfer. Embryo splitting is performed by surgical bisection of early tubal stage embryos, while somatic cell nuclear transfer involves fusing the nucleus of a single diploid cell with an enucleated and unfertilized ovum (Boa-Amponsem and Minozzi, 2006). Embryonic stem cells are derived from the totipotent cells of the early embryo and can give rise to all differentiated cells, including germ line cells.

C. Utilization of GRFA

Genetic resources are the raw material for agricultural development and for the continued survival of natural populations. Therefore their sustainable use is crucial for global food security and economic well-being. Biotechnologies are increasingly being applied for the enhancement of GRFA and have had a profound impact on their effective utilization.

1. Reproductive biotechnologies (livestock and fisheries/aquaculture)

In addition to the reproductive technologies described earlier, sperm and embryo sexing permits the preferential production of one sex in livestock.

a. Sperm sexing

Depending upon the species, the X chromosome contains 2–5% more DNA than the Y chromosome (Boa-Amponsem and Minozzi, 2006). The DNA contained in the sperm has distinct emission patterns when stained with a fluorescent dye and exposed to light, thus allowing separation by flow cytometry.[44]

b. Embryo sexing

Y chromosome-specific DNA probes, karyotyping, male-antigen specific antibodies, and X-linked activity enzymes are used for embryo sexing, with Y chromosome-specific DNA probes being the most reliable.

c. Hormonal treatment

Similar to livestock, where hormonal treatment is applied for estrus synchronization, in aquaculture hormonal treatment is used for two main purposes, that is, controlling the time of reproduction and developing monosex populations. Chemically synthesized hormones are relatively inexpensive and practical to use.

[44]Automated measurements on large numbers of individual cells or other small biological materials, made as the cells flow one by one in a fluid stream past optical and/or electronic sensors (FAO, 2001).

2. Biotechnologies for disease diagnostics (cross-sectoral)

Diseases are a major impediment to the sustainable utilization of GRFA. Biotechnologies, based on immunoassays and on nucleic acid detection, for pathogen screening and disease diagnostics are important in all agricultural sectors and can contribute to improved plant and animal disease control and to food safety.

ELISA can be applied either for the detection of an antigen (that is, pathogen) or for the detection of antibodies produced by the host in response to the pathogen. Both monoclonal[45] and polyclonal[46] antibodies can be employed for the assay. The development of recombinant antigens has further improved the rapidity, specificity, sensitivity, and safety of ELISA. Competitive ELISA (C-ELISA) detects the antibody or antigen based on competition between the test serum and the detecting antibody. This method has an added advantage since a relatively crude/impure sample can be used for the assay (OIE, 2009).

PCR is a highly sensitive procedure and the most widely applied molecular technique for disease diagnostics. It allows the amplification of a specific DNA sequence from a complex mixture of heterogeneous sequences. The amplified DNA can then be identified using gel electrophoresis or hybridization with a labeled DNA probe. Since PCR detects pathogen DNA rather than host antibodies, it provides more rapid results by eliminating the lag period between the initial infection and the appearance of detectable amounts of antibodies.

In the case of certain viruses, whose genomes are made of RNA, reverse transcriptase PCR (RT-PCR) is necessary, that is, a complementary DNA (cDNA) copy is first synthesized using reverse transcriptase which then acts as the template for amplification. The sensitivity of PCR can be further enhanced by utilizing a technique referred to as nested PCR, that is, a second set of primers is used to amplify a subfragment of the first PCR product. Real-time/quantitative PCR provides quantitative information by measuring the accumulation of the PCR product during the amplification reaction. It is more rapid than conventional PCR techniques, thus reducing the risk of cross-contamination, and can be scaled up for high-throughput applications. The loop-mediated isothermal amplification (LAMP) PCR is a robust cost–effective test that is performed at one temperature (that is, no need for expensive temperature ramping PCR machines) and the amplified product can be detected visually.

RFLPs can be used to distinguish between isolates of closely related pathogens. When integrated with PCR (PCR-RFLP), it offers improved sensitivity and is particularly valuable when the pathogen is available in small

[45]An antibody, produced by a hybridoma, directed against a single antigenic determinant of an antigen (FAO, 2001).

[46]A serum sample that contains a mixture of distinct immunoglobulin molecules, each recognizing a different antigenic determinant of a given antigen (FAO, 2001).

numbers or is difficult to culture. Microarrays may also be used to detect pathogens, especially when more than one pathogen might be present, although more extensive application of this technology for disease diagnosis would require it to become less expensive.

3. Biotechnologies for disease control (livestock and fisheries/aquaculture)

Biotechnologies have also been extensively used in the development of vaccines for preventing and thereby managing diseases in livestock and fish. A vaccine is a preparation of dead or attenuated (weakened) pathogens, or of derived antigenic determinants, that can induce the formation of antibodies in a host and thereby produce host immunity against the pathogen. Molecular techniques can facilitate the identification of potential antigen candidates that may be effective in vaccines (e.g., by using monoclonal antibodies and expression libraries), construction of new candidate vaccines (e.g., by using PCR and cloning) as well as assessment of candidate vaccine efficacy, its mode of action, and host response (e.g., by using quantitative RT-PCR) (Kurath, 2008).

The different types of vaccines, produced by molecular methods, are described below (more details available in OIE, 2010a).

a. Gene deletion vaccines

Gene-deleted pathogens, with deletions in genes associated with virulence or involved in key metabolic pathways, can be used as live vaccines as they retain the immunogenicity of the wild-type organism but have reduced virulence. Live attenuated vaccine strains of bacteria have been created that confer better protection than killed vaccines.

b. Virus-vectored vaccines

Suitable vectors are essential for the efficient delivery of protective antigens into the animal. Many viruses have been used as vaccine delivery vehicles since their genomes can accommodate large amounts of exogenous DNA. Replication competent as well as replication defective vectors have been commercially developed. Furthermore, a virus may act both as a vector and as a self-vaccine (e.g., the recombinant capripox virus expressing a peste des petits ruminants virus, antigen).

c. DNA vaccines

Recombinant DNA technology has made possible the construction of safer and more cost–effective vaccines since only the desired antigen (that is, unable to replicate or induce disease) instead of the entire pathogen is used. DNA vaccination involves the direct inoculation, into the animal, of an antigen-encoding bacterial plasmid to elicit the immune response. DNA vaccines can stimulate the induction of both humoral- and cell-mediated immune responses, vital for protection against a wide range of pathogens. They are relatively easy to produce in addition to being very stable with a long shelf life. Additionally, the need for complex vector organisms is obviated.

d. Subunit vaccines

Subunit vaccines are composed of protein or glycoprotein components of a pathogen that are capable of inducing a protective immune response and may be produced by conventional biochemical or recombinant DNA technologies. Recombinant subunit vaccines have distinct advantages over live attenuated and inactivated vaccines since they are efficient in inducing humoral- and cell-mediated immunological responses, and the risks associated with handling the pathogen are eliminated. However, subunit vaccines may be more expensive and may require specific adjuvants[47] to enhance the immune response.

e. Marker vaccines

In addition to limiting the clinical impact of the disease, marker vaccines optimize the effectiveness of the vaccination strategy. Such vaccines, together with companion diagnostic tests, are of paramount importance for evaluating disease eradication programs since they are capable of differentiating infected and vaccinated animals (DIVA) in case of an outbreak. DIVA vaccines have at least one less antigenic protein than the corresponding wild-type virus and are based on detecting the serological response either toward a protein whose gene has been deleted in the vaccine strain or against virus nonstructural proteins not present in subunit vaccines and highly purified vaccines.

4. Chromosome set manipulation (crops, forest trees, and fisheries/aquaculture)

Chromosome set manipulation, that is, alteration of the chromosome set, is used for a range of different purposes in agriculture. In fish and plants, this technique has been used to induce polyploidy (to create sterile individuals) and for uniparental chromosome inheritance (valuable in breeding programs and for the establishment of monosex populations).

[47]An adjuvant is a substance that enhances immune responses when coadministered with the antigen.

5. Tissue culture-based techniques (crops and forest trees)

FAO studies on the development, adoption, and application of biotechnologies in developing countries indicate that tissue culture is the most common biotechnology technique for plant genetic resources, being applied in 88% of the 25 developing countries surveyed (Sonnino *et al.*, 2007). Similarly, FAO (2004) highlights its importance for the forestry sector, where micropropagation activities were reported in at least 64 countries worldwide, mainly in Asia, Europe, and North America.

a. Wide crossing

Interspecific hybridization/wide crossing involves crossing plants belonging to two different species that are not normally sexually compatible. It is used to transfer useful characteristics from wild relatives to cultivated species or to combine favorable traits of two different species, but significant amounts of time and scientific expertise need to be invested. Biotechnology approaches such as *in vitro* embryo rescue and anther culture are crucial to overcome sexual incompatibility and to speed up the process. *In vitro* embryo rescue utilizes a sequence of tissue culture techniques to enable a fertilized immature embryo, resulting from an interspecific cross, to avoid abortion caused by unbalanced endosperms and to continue growth and development until it can be regenerated into an adult plant. Anther culture is the aseptic culture of immature anthers (within which pollen develops and matures) to generate haploid plants. The chromosome number is then doubled through the application of chemicals such as colchicine or other *in vitro* techniques. Colchicine is also used to artificially double the chromosome number of sterile wild hybrid plants in order for functional pollen and eggs to be produced and allopolyploid[48] fertile progenies to be obtained.

b. Somatic hybridization

Another method to circumvent reproductive barriers and introduce novel genes into a plant's genome from a dissimilar donor species is somatic hybridization. It involves the induced *in vitro* fusion of protoplasts or cells of two genetically different parents and the subsequent regeneration of adult plants from the resultant hybrid cells. Application of somatic hybridization is limited by the regeneration rate of the hybridized cells, which can be very low for certain hybrid combinations.

c. Micropropagation

Micropropagation refers to the *in vitro* multiplication and/or regeneration of plant material under aseptic and controlled environmental conditions to produce thousands or millions of plants for transfer to the field. It is a fast- and low-cost method to

[48]A polyploid organism with sets of chromosomes derived from different species (FAO, 2001).

overcome the accumulation of infectious agents in vegetatively propagated plants and has been used for mass clonal propagation of true-to-type, disease-free material in more than 30 developing and transition countries (Sonnino *et al.*, 2009).

6. Molecular marker-assisted selection (cross-sectoral)

An alternative or complement to conventional phenotypic selection is molecular marker-assisted selection (MAS), where the desired trait is selected indirectly by selecting for a marker(s) genetically linked to a gene or genes influencing it. MAS can thus greatly accelerate genetic improvement by enhancing the accuracy of selection and reducing the time needed (particularly when phenotype screening is difficult or when the trait is expressed late in the life of the individual). A successful MAS strategy is dependent upon the availability of a genetic map with sufficient number of regularly spaced markers, tight linkage between the gene or QTL of interest and adjacent markers, adequate recombination between the markers and the rest of the genome as well as the ability to analyze a larger number of individuals in a time- and cost–effective manner. Furthermore, MAS is best utilized when embedded within existing conventional breeding programs that already have performance and pedigree recording systems in place.

Genomics tools for the simultaneous screening of tens of thousands of genetic markers in the form of SNPs have allowed for the application of so-called genomic selection (GS), an advanced form of MAS. In this procedure, a training set of data consisting of genomic and phenotypic information on a subset of the population is used to establish statistical associations between markers and phenotypes and to develop equations for prediction of breeding values that are then applied to individuals without phenotypic data, that is, the validation set (Meuwissen *et al.*, 2001). GS does not require *a priori* phenotypic information for the validation set and thus can increase the genetic gain by accelerating the breeding cycle.

Even though genotyping is becoming less expensive than phenotyping in applied breeding programs (due to an increase in the number of markers and reduced cost per data point), GS is not yet applied in developing countries. One reason is that the accuracy of breeding value estimation depends on the number of individuals in the reference population with which to construct the training set (the population should also be genetically and phenotypically characterized), and considerable financial and technical resources are required to establish such reference populations. Moreover, using reference data from developed country populations for selection is complicated due to differences in breeding objectives between the production systems.

7. Fermentation (microbial)

Fermentation is the anaerobic breakdown of complex organic substances into simpler substances by microbes (FAO, 2001). In addition to extending the shelf-life, quality and safety of food, it is widely applied to produce a variety of metabolites including vitamins, antimicrobial compounds, enzymes, flavors, fragrances, food additives, and a range of other high value-added products.

8. Biofertilizers (microbial)

Biofertilizers are preparations containing live or latent cells of agriculturally beneficial strains of microorganisms that are applied to seed or soil to build up the numbers of such microorganisms and accelerate certain microbial processes to augment nutrient acquisition by plants (Motsara and Roy, 2008). Biofertilizers consist of nitrogen fixers (*Rhizobium*, *Azotobacter*, *Azospirillum*, cyanobacteria/blue-green algae, *Azolla*), phosphate solubilizing microorganisms, and mycorrhizal fungi.

9. Biopesticides (microbial)

Biopesticides are living organisms or natural products derived from these organisms that are mass produced and employed to control pests (Bailey *et al.*, 2010; Chandler *et al.*, 2008). The organisms employed can be insects, nematodes, or microorganisms, while naturally occurring substances include plant extracts and insect pheromones. Microorganisms are the commonest biopesticides and include protozoa, bacteria, fungi, and viruses, and of these, the most dominant is *Bacillus thuringiensis* (Chandler *et al.*, 2008).

10. Bioremediation (microbial)

Bioremediation refers to the use of living organisms to remove contaminants, pollutants, or unwanted substances from soil or water (FAO, 2001). This technology mainly utilizes microbes, although the cultivation of plants can accelerate bioremediation since the rhizosphere[49] provides a favorable environment for microbial proliferation (Stout and Nüsslein, 2010; Wenzel, 2009). Such plant-assisted bioremediation is referred to as phytoremediation and can also be accomplished by exploiting plant–endophyte[50] partnerships (Weyens *et al.*, 2009). In aquaculture, integrated multitrophic aquaculture strategies have been developed for bioremediation that employ complementary organisms such as bacteria, seaweeds, shellfish, filter feeders, and bottom feeders to optimize nutrient utilization and waste treatment (Chávez-Crooker and Obreque-Contreras, 2010).

[49]The soil region in the immediate vicinity of growing plant roots.
[50]A microorganism that lives inside a plant.

11. Probiotics (microbial)

Probiotics are live microorganisms which, when administered in adequate amounts, confer a health benefit on the host (FAO/WHO, 2001). Foods containing probiotics are considered to be important in human health and nutrition, with lactic acid bacteria reported to have the most beneficial effects in the human gastrointestinal tract (Burgain *et al.*, 2011). Probiotics are also included as feed additives for both livestock and fish, with strains of *Lactobacillus* and *Bifidobacterium* generally used for monogastric animals, and *Aspergillus oryzae* and *Saccharomyces cerevisiae* for ruminants (FAO, 2011c), while *Lactobacillus*, *Carnobacterium*, *Vibrio*, *Bacillus*, and *Pseudomonas* are prevalent in finfish and shellfish aquaculture (Balcazar *et al.*, 2006). Probiotics are beneficial in multiple ways, such as conferring antagonism against pathogens, strengthening the host's immune system, and providing nutritional benefits (Soccol *et al.*, 2010). Their efficacy can be increased by supplementation with prebiotics, that is, nondigestible oligosaccharides (Nayak, 2010a).

12. Mutagenesis (cross-sectoral)

Chemical, radiation, or somaclonal mutagenesis can be used to accelerate the process of spontaneous mutation to create novel phenotypes. Mutagenesis is one of the few biotechnologies that is employed more in developing countries than elsewhere, with the FAO/International Atomic Energy Agency (IAEA) partnership being instrumental in technology transfer of mutation breeding approaches.

13. Transgenesis (cross-sectoral)

Transgenesis/genetic modification refers to the introduction of exogenous DNA or RNA sequences into an organism's genetic material to create a genetically modified organism (GMO). The input gene(s) may be from a different kingdom, a different species within the same kingdom or even from the same species (FAO, 2009a). Genetic modification has been at the center of a highly polarized debate worldwide in recent years, due to which non-GMO biotechnologies, and their potential benefits for food security and sustainable development in developing countries, have tended to be neglected in most discussions about biotechnology, often hindering their development and application. This is despite the fact that the major breeding and genetic resource management applications to date have come from non-GMO biotechnologies (FAO, 2011a–e). For these reasons, and because of the already very extensive literature regarding the many different issues surrounding their use, GMOs are not discussed in detail in this chapter.

III. CURRENT STATUS OF BIOTECHNOLOGIES FOR THE MANAGEMENT OF CROP GENETIC RESOURCES

Estimates indicate that an additional 1 billion tons of cereals will need to be produced annually by 2050 in order to meet the needs of the world's population (FAO, 2009b). It is projected that the majority of this increase in crop production will come from intensification and a small percentage from the expansion of arable land. However, the rate of growth in yields of the major cereal crops has been steadily declining globally; an estimated 75% of crop diversity has been lost between 1900 and 2000 (FAO, 2010c), and 16–22% of the wild relatives of important food crops of peanut, potato, and beans are predicted to disappear by 2055 due to the impact of climate change (Jarvis *et al.*, 2008).

Consequently, there is a need for addressing productivity constraints in conjunction with conserving the natural resources base, while addressing the need for improving farming sustainability and facing the potential challenges posed to crops and agrosystems by climate change. Productivity constraints may also be addressed through the focused promotion of noncereal crops which could benefit the most from future variety improvement because their yields are still far from the theoretical crop productivity potential. Noncereal crops are also less affected than grains by global price fluctuations, which places them in a favorable position to address food security in the poorest countries. Another critical factor underpinning the future of food security is the conservation and utilization of crop wild relatives (CWR)[51] populations, since they contain far more genetic diversity than the crops themselves. As reviewed in this section, rapid scientific advances in crop biotechnologies, especially in the past two decades, have provided important tools for tapping into the diversity of CWR and for enhancing the management of crop genetic resources, for example, through investigating genetic diversity, developing more effective conservation strategies, and obtaining improved crop varieties.

A. Molecular markers

Knowledge of the genetic variation within crops and their phylogenetic relationships with wild relatives is important for informed decision-making and designing appropriate conservation and breeding strategies. Molecular markers have been employed for genetic diversity studies in many crop species, including wheat (Hai *et al.*, 2007; Zarkti *et al.*, 2010), rice (Huang *et al.*, 2010a; Saker *et al.*, 2005), maize (Lanteri and Barcaccia, 2006; Van Inghelandt *et al.*, 2010), barley (Orabi *et al.*, 2009; Wang *et al.*, 2010a), common bean (Blair *et al.*, 2010; Jose *et al.*, 2009), sorghum (Ali *et al.*, 2008; Pei *et al.*, 2010), sugarcane (Kawar *et al.*, 2009; Singh *et al.*,

[51]The wild ancestors of crop plants and other species closely related to crops.

2010), and potato (Akkale *et al.*, 2010; Fu *et al.*, 2009). Markers have provided information on the ecological or geographic patterns of diversity distribution in numerous crops and their wild relatives (Rao, 2004). The use of molecular markers for genetic diversity studies in crop species has increased worldwide in the past decade although it is still limited in developing countries (FAO, 2010c).

Molecular markers have also been used to evaluate the effectiveness of different conservation strategies on the genetic structure of populations (Lanteri and Barcaccia, 2006). Marker analysis of a threatened common bean landrace showed that *in situ* conservation is the most effective way to maintain the diversity (Negri and Tiranti, 2010), while for capsicum, novel genetic variation was found in both the *in situ* population studied and some *ex situ* accessions, thus supporting conservation of this species via both strategies (Votava *et al.*, 2002). The genetic diversity of maize populations maintained *in situ* and *ex situ* was found to be substantially equal (Rice *et al.*, 2006).

Determination of the impact of adoption of improved varieties on the genetic diversity of germplasm is essential to assess the need for further incorporation of exotic germplasm into the existing breeding pool. Microsatellite studies with Nordic spring wheat cultivars illustrated that genetic diversity was enhanced by breeders in the first quarter of the twentieth century, followed by a decrease and then again an increase during the second quarter of the century (Christiansen *et al.*, 2002). Similar results were reported in other studies assessing the diversity of bread wheat in the United Kingdom (Donini *et al.*, 2000) and of durum wheat (Maccaferri *et al.*, 2003). Molecular investigations with rice cultivars have revealed that while genetic diversity has increased over time in Italy (Mantegazza *et al.*, 2008), it has declined to a certain extent in recent decades in China (Wei *et al.*, 2009).

Molecular markers have provided insight into the identification of crop progenitors, origins of domestication, and the molecular changes underlying domestication traits (Burger *et al.*, 2008; Gross and Olsen, 2010). For example, microsatellite data proved that maize is the product of a single domestication event from its wild progenitor in southern Mexico (Matsuoka *et al.*, 2002) as is einkorn wheat from its wild relatives in south-east Turkey (Heun *et al.*, 1997) and pearl millet in West Africa (Oumar *et al.*, 2008). In contrast, multiple domestication events have been uncovered for barley (Morrell and Clegg, 2007), common bean (Chacón *et al.*, 2005), and squash (Sanjur *et al.*, 2002). To date, nine domestication genes have been identified in plants in addition to 26 other loci that underlie crop diversity associated with human cultural preferences or different agricultural environments (Purugganan and Fuller, 2009).

Introgression between wild populations and cultivated plants is a widespread phenomenon with 12 of the 13 most important food crops of the world hybridizing with wild relatives in some regions of their agricultural distribution

(Ellstrand *et al.*, 1999). Overwhelming gene flow from crops can deplete the genetic diversity of wild populations, leading in some cases to their genetic extinction, while crop-to-weed gene flow has the potential to promote the evolution of more aggressive weeds (Andersson and de Vicente, 2010; Papa, 2005). Gene flow is particularly difficult to assess when crops and wild species are very closely related. In such cases, molecular markers have proved to be informative tools to monitor gene flow from cultivated crops and ponder over its consequences, for example, for rice (Chen *et al.*, 2004) and sorghum (Morrell *et al.*, 2005) with respect to their wild and weedy relatives. Marker analysis with common bean demonstrated that gene flow was about three- to fourfold higher from domesticated to wild populations than in the reverse direction. Further, the weedy populations had intermediate traits between wild and domesticated bean plants suggesting that they were hybrids, rather than escapes from cultivation (Papa and Gepts, 2003).

Considerable progress has been made in identifying and consequently utilizing QTLs for breeding in crop species. Molecular markers tightly linked with many agronomic, stress tolerance, and disease resistance QTLs are available in major crop species (Bernardo, 2008; Collins *et al.*, 2008; Yadav *et al.*, 2011a) and their wild relatives (Swamy and Sarla, 2008). Genetic linkage maps of cultivated and wild *Vigna* crop species (cowpea, mung bean, rice bean, azuki bean, and black gram) have been developed and used to identify QTLs of domestication-related traits (Takeya *et al.*, 2011).

QTL mapping experiments are generally heterogeneous, and hence comparative QTL mapping is necessary to synthesize all the information for MAS purposes. For a few species (rice, wheat, maize, barley, oat, sorghum, pearl millet, foxtail millet, and wild rice), the Gramene QTL database integrates results from independent experiments for 11,624 QTLs for numerous traits.[52] For other species, QTL meta-analysis has been performed for traits of interest to underscore chromosomal regions for use in breeding (Danan *et al.*, 2011; Lanaud *et al.*, 2009).

Molecular markers are of considerable value in first identifying populations for collecting by developing optimum sampling strategies for gene banks and, subsequently, for the management of the accessions held. For example, AFLP marker studies undertaken to evaluate Sri Lankan coconut populations showed that the greatest diversity is found within populations rather than between populations, and hence emphasis should be placed on collecting a large number of plants from a few populations (Perera *et al.*, 1998).

Molecular markers assist in the management of conserved germplasm by increasing the efficiency of gene bank operations. They are an effective tool to first characterize accessions in the gene bank and help identify useful traits.

[52]http://www.gramene.org/qtl/.

Molecular markers can identify both gaps (missing/underrepresented populations) and redundancies (duplicate accessions as opposed to safety duplicates) in collections to guide future acquisition and increase cost–effectiveness.

Currently, less than 30% of the 7.4 million plant germplasm accessions held in gene banks worldwide are estimated to be sufficiently distinct (FAO, 2010c). It has been calculated that the additional cost of identifying a duplicate cassava accession (that is to be added to the collection) using molecular characterization, once passport data has been checked, is about 12 times less than the cost of conserving and distributing the material as a different accession in-perpetuity (Horna *et al.*, 2010).

The International Potato Center (CIP) in Peru employed molecular markers to compare accessions of sweet potato that appeared identical morphologically, thus identifying duplicates and reducing their clonal collections by approximately two-thirds (Dawson *et al.*, 2009). However, in AFLP studies with a wild potato collection, the costs to detect redundancies were estimated to be approximately 2.5 times higher than the savings expected per generation by the reduction of the collection (Van Treuren *et al.*, 2004). Thus, the benefits of a reduced collection do not always offset the requisite investments to identify redundancies, which vary depending on the number/type of markers employed, the regeneration costs of the concerned crop, the final level of redundancy detected, and the timeframe within which investment returns are expected (Van Treuren *et al.*, 2010). Nevertheless, it must be mentioned that progress in marker technologies is proceeding at a fast pace and costs are dropping, albeit not at the same rates witnessed in DNA sequencing technologies. In such a changing landscape, it can be envisioned that markers will play an increasingly important role in the management of gene banks.

Periodic regeneration and multiplication are essential features of gene bank management (to maintain viability and replenish stocks for distribution), following which assessment of genetic integrity is crucial. The size of the seed sample for regeneration should reflect both the reproductive biology of the species under consideration and the degree of homogeneity/heterogeneity of the accession.[53] In this respect, molecular markers can assist in estimating the effective population size (N_e) that will ensure the genetic integrity of the accessions. Molecular markers can also be employed to verify accession identity, detect inadvertent seed mixtures, and monitor changes in alleles/allele frequencies as well as gene flow between accessions (de Vicente *et al.*, 2006; Spooner *et al.*, 2005).

A study investigating the genetic identity of wheat accessions (regenerated 24 times), using microsatellites, did not detect any unintended pollen or seed mixing for any of the accessions but found that genetic drift had

[53]http://typo3.fao.org/fileadmin/templates/agphome/documents/PGR/ITWG/ITWG5/ITWG5_INF3Upton.pdf.

occurred in one case (Börner *et al.*, 2000). Another study with AFLPs, comparing wheat accessions (derived from identical sources) duplicated at two different gene banks, concluded that while the overall genetic diversity was conserved at both locations (compared to the original collections preserved without regeneration) there was possible unintentional selection at one gene bank (Hirano *et al.*, 2009). Therefore, efficient regeneration strategies coupled with continual monitoring of the genetic diversity conserved in gene banks are imperative.

Molecular markers can also be used to make larger collections more accessible and useful for allele mining by developing core collections, that is, subsets that consist of a small percentage of the entire collection while still representing a broad spectrum of genetic variability. Markers have generated genetic diversity information to assist in the establishment of core collections for many species, including maize (Franco *et al.*, 2006; Qi-Lun *et al.*, 2008), wheat (Hao *et al.*, 2006), rice (Ebana *et al.*, 2008), potato (Ghislain *et al.*, 2006), chick pea (Upadhyaya *et al.*, 2008), grape (Le Cunff *et al.*, 2008), cacao, and pepper (Marita *et al.*, 2000). For creating core collections, emphasis should be placed on methodologies that use data generated by markers in concert with the morphological and agronomical characterization of the accessions (Balfourier *et al.*, 2007; Jansen and van Hintum, 2007).

Core collections for certain species, particularly grasses, are still sizeable and therefore of limited use to breeders. Molecular markers have been used to develop thematic core collections that are much smaller in size but exhibit the maximum allelic richness for specific traits of interest (Pessoa-Filho *et al.*, 2010). A germplasm collection, with a defined core collection together with various thematic core collections focusing on different traits, thus has the potential to increase the efficiency of germplasm use in breeding programs.

The establishment and maintenance of gene banks must be coupled with the ability to identify useful genes and utilize the genetic diversity with much greater efficiency. To facilitate the generation and exchange of standardized molecular marker data for plant germplasm held in gene banks, a list of descriptors (de Vicente *et al.*, 2004) and guidelines for developing new descriptor lists (Bioversity International, 2007) have been produced although, in general, the amount of characterization data is quite low. The situation is further exacerbated in developing countries, where the percentage of accessions characterized using molecular markers is less than 12%, with the exception being 64% in the Near East (FAO, 2010c). This lack of adequate characterization is a major impediment to the sustainable use of GRFA (even though the number of accessions deposited is continuously growing), hence characterizing the extensive collections maintained in gene banks must be prioritized.

An additional challenge, especially in developing countries, is the management of the molecular, phenotypic, and agricultural data that is being generated across the gene banks. The Germplasm Resource Information System Global Project,[54] a partnership between the Global Crop Diversity Trust, Bioversity International and the Agricultural Research Service of the United States Department of Agriculture, is being developed to provide the world's crop gene banks with a powerful, but easy-to-use plant genetic resource information management system. It will also allow researchers and breeders to access and utilize the information more effectively.

B. "Omics"

The genomes of several plant species have been sequenced to date, including rice, maize wheat, sorghum, soybean, pigeonpea, cassava, potato, and tomato (Mochida and Shinozaki, 2010; Varshney *et al.*, 2012). Whole genome sequences have led to the discovery of thousands of SNPs to create high-density genetic maps (Ganal *et al.*, 2009) and, together with functional "omic" technologies, have greatly facilitated identification of candidate genes (Langridge and Fleury, 2011). For example, transcriptome profiling in concert with QTL mapping has been used to detect novel candidate genes associated with tuber quality traits in potato (Kloosterman *et al.*, 2010) and grain number in rice (Deshmukh *et al.*, 2010).

Next-generation sequencing (NGS) technologies allow DNA sequencing at a much higher speed and greatly reduced cost and are being employed for a range of applications (Deschamps and Campbell, 2010; Varshney *et al.*, 2009a). Markers can be discovered on a genome-wide scale by aligning short reads of sequence from the genotype of interest to a reference genome (Huang *et al.*, 2009a) or through the *de novo* sequencing of species without reference sequences (Bundock *et al.*, 2009). In polyploid species with complex genomes, SNPs have been detected after the creation of cDNA libraries to avoid sequencing repetitive regions (Barbazuk *et al.*, 2007). Transcriptome sequencing has also been useful for less-characterized species such as chick pea, for the identification of SNPs.[55]

A technology that has recently gained attention for genetic characterization is a hybridization-based method, diversity arrays technology (DArT). Diversity panels are created using DNA fragments pooled from various varieties/cultivars which are then hybridized with individual DNA samples. Differential hybridization reveals polymorphisms between the samples (Jaccoud *et al.*, 2001). DArT simultaneously assays thousands of markers in parallel across samples, and since it does not require prior DNA sequence information, it is of special interest to orphan crops. It is a low-cost high-throughput system,

[54]http://www.grin-global.org.
[55]http://www.intl-pag.org/16/abstracts/PAG16_P05f_385.html.

although a prerequisite is the development and validation of a diagnostic DArT array. This technology has been applied successfully in many species, including wheat (Akbari *et al.*, 2006), rice (Xie *et al.*, 2006), cassava (Xia *et al.*, 2005), barley (Wenzl *et al.*, 2004), pigeon pea (Yang *et al.*, 2006), sorghum (Mace *et al.*, 2009), and oat (Tinker *et al.*, 2009).

Genome-wide scans for selection signatures related to domestication have been carried out in only a small number of crop species (Chapman *et al.*, 2008; Wright *et al.*, 2005). In rice, it has been suggested that selection could leave a genome-wide imprint, rather than a localized signature (Caicedo *et al.*, 2007). Genome-wide association mapping studies, correlating phenotype data with genome-wide genotypes, are beginning to be used to detect loci for agronomically important traits in a few crops like wheat (Neumann *et al.*, 2011), maize (Tian *et al.*, 2011), barley (Cockram *et al.*, 2010), and rice (Huang *et al.*, 2010b). However, this technique has lower sensitivity for detecting rare alleles and can be ineffective if there is strong genetic differentiation within populations (Nordborg and Weigel, 2008).

Comparative genomics has contributed to understanding genetic diversity, offered insights into evolution, and enhanced gene discovery (Duran *et al.*, 2008; Flavell, 2008). The CGIAR Generation Challenge Program (GCP)[56] used this approach to identify orthologous genes[57] for improving cereal yields for maize, rice, and sorghum in high-aluminum and low-phosphorous soils.

Transcriptomic approaches have been used to characterize crop species and to improve the understanding of complex responses, for example, by evaluating global gene expression changes in response to disease-causing pathogens (Sana *et al.*, 2010), symbiotic associations (Hocher *et al.*, 2011), abiotic stresses (Narsai *et al.*, 2010), low phosphorus (Li *et al.*, 2010a), and aluminum phytotoxicity (Mattiello *et al.*, 2010). Evolving proteomic technologies have facilitated the investigation of developmental processes (Agrawal and Rakwal, 2006), abiotic stress tolerance (Manaa *et al.*, 2011; Sobhanian *et al.*, 2011), and the detection of plant pathogens as well as study of plant–microbe interactions (Kav *et al.*, 2007).

Metabolomics is also emerging as a promising tool for the fundamental biochemical comprehension of plant metabolism, analysis, and discovery of a broad range of metabolites; for determining metabolite relationships with specific quality traits; and for discriminating taxonomic relationships (Fernie and Schauer, 2009; Guy *et al.*, 2008; Shepherd *et al.*, 2011; Summer, 2010).

[56]http://www.generationcp.org/.

[57]These are genes in different species that can be traced back to the same common ancestor and normally retain the same function in the course of evolution.

C. Bioinformatics

Sequencing is becoming simpler and cheaper, and several bioinformatic tools are available for allele mining analysis of these data, such as PLACE[58] and plant-Care[59] (Kumar *et al.*, 2010). Providers for plant genome sequences and annotations that also facilitate comparative genomic studies include PlantGDB[60] with sequence data for over 70,000 plant species; Phytozome[61] with access to 26 species; Rice Genome Annotation Project[62]; MaizeGDB[63]; SoyBase[64]; Brassica Genome Gateway[65]; SOL genomics network[66] for Solanaceae species; GrainGenes[67] for wheat, barley, rye, and oat; and Gramene[68] for grasses. Plant-specific proteomic databases include the rice proteome database[69] and the soybean proteome database.[70] Metabolic platforms include the Plant Metabolic Network,[71] the Metabolome Tomato Database,[72] and the Armec Repository Project[73] for potato. More details on bioinformatic resources for plants can be found in Skuse and Du (2008).

D. Cryopreservation

Cryopreservation has been applied to over 200 plant species using diverse materials such as seeds, cell suspensions, callus cultures, meristematic tissue, pollen, and somatic and zygotic embryos (Dulloo *et al.*, 2010; Harding, 2010). It is particularly important for the long-term storage of vegetatively propagated crop species such as cassava and banana, as well as species that produce recalcitrant seeds[74] such as coconut and mango. Generally, the plants recovered after cryopreservation maintain their genetic integrity and are true-to-type (Harding,

[58]http://www.dna.affrc.go.jp/PLACE/.
[59]http://bioinformatics.psb.ugent.be/webtools/plantcare/html/.
[60]http://www.plantgdb.org/.
[61]http://www.phytozome.net/.
[62]http://rice.plantbiology.msu.edu/.
[63]http://www.maizegdb.org/.
[64]http://www.soybase.org/.
[65]http://brassica.bbsrc.ac.uk/.
[66]http://solgenomics.net/.
[67]http://wheat.pw.usda.gov/GG2/index.shtml.
[68]http://www.gramene.org/.
[69]http://gene64.dna.affrc.go.jp/RPD/.
[70]http://proteome.dc.affrc.go.jp/cgi-bin/2d/2d_view_map.cgi.
[71]http://www.plantcyc.org/.
[72]http://appliedbioinformatics.wur.nl/moto/.
[73]http://www.armec.org/MetaboliteLibrary/index.jsp.
[74]Recalcitrant seeds are seeds that unable to germinate after cold storage and/or desiccation.

2004; Liu *et al.*, 2008; Zarghami *et al.*, 2008), although, in some cases, variation among genotypes of the same species in DNA structure and methylation patterns have been observed (Johnston *et al.*, 2009; Kaity *et al.*, 2008).

Orthodox seeds display natural tolerance to desiccation and cryogenic temperature stresses and thus can be cryopreserved without any pretreatment. However, most hydrated tissues are highly sensitive to freezing injury. Cryopreservation is then carried out employing classical freeze-induced dehydration techniques[75] or newer vitrification techniques.[76] Vitrification is the most extensively used cryopreservation technique since it has higher reproducibility and is more appropriate for complex organs like shoot tips and embryos (Reed, 2008).

Cryopreservation protocols are more advanced for vegetatively propagated species, including for varieties within a given species, and numerous plantation crops, fruit trees as well as roots and tubers have been successfully cryopreserved. The Global Crop Diversity Trust is supporting work on the development and refinement of robust cryopreservation protocols for yam, sweet potato, and aroids, and specific genotypes of cassava that are not responding to existing protocols.[77] In contrast, protocols for species that produce recalcitrant seeds are less advanced owing to their seed characteristics, such as high desiccation sensitivity and structural complexity (Engelmann, 2011).

The choice of genetic material for cryopreservation depends on the conservation goal. Cell suspensions and callus cultures are cryopreserved in order to conserve their specific features that might be lost during routine *in vitro* maintenance. For example, rice calli stored in liquid nitrogen exhibit a higher competence for transformation compared to their unfrozen counterparts (Moukadiri *et al.*, 1999). Pollen is cryopreserved for use in breeding programs, preserving nuclear genes and investigating fundamental aspects of pollen biology (Towill and Walters, 2000). Pollen from 600 accessions belonging to 40 species has been cryopreserved in India (Ganeshan and Rajashekaran, 2000), while pollen from more than 700 accessions is cryopreserved in China (Li *et al.*, 2009a).

Shoot meristematic tissue is the most commonly used explant for long-term storage of vegetatively propagated species. Over 1000 old potato varieties are cryopreserved at the Institute of Plant Genetics and Crop Plant Research in Germany (Keller *et al.*, 2006), 345 potato accessions at the CIP (Panis and

[75]Slow cooling to a defined temperature in the presence of a cryoprotectant, followed by rapid immersion in liquid nitrogen.

[76]Cell dehydration is performed prior to freezing in the presence of cryoprotective media, followed by rapid cooling.

[77]http://www.croptrust.org/.

Lambardi, 2006), 540 cassava accessions at the International Center for Tropical Agriculture (CIAT) (Gonzalez-Arnao *et al.*, 2008), and 630 banana accessions at the International Network for the Improvement of Banana and Plantain (INIBAP) International Transit Center (Panis *et al.*, 2007). In the Republic of Korea, two garlic cryocollections with more than 800 accessions have recently been established (Kim *et al.*, 2009).

Cryopreservation is also applied to orthodox seeds, especially of rare and endangered species, to extend seed longevity (Mandal, 2000; Touchell and Dixon, 1994). For example, protocols have been developed to cryopreserve whole seeds of *Coffea arabica* L., which obviate the necessity of germinating seeds or embryos *in vitro* after cryopreservation (Dussert and Engelmann, 2006). Coffee seeds can be rewarmed after cryopreservation and sown directly in vermiculite in the greenhouse. Further, aging time course studies for lettuce seeds stored at temperatures between 50 and −196 °C have shown that cryopreservation can prolong the shelf-life of orthodox species, compared to storage under conventional optimal conditions of low temperature and low moisture content (Dulloo *et al.*, 2010; Walters *et al.*, 2004).

Cryopreservation may have practical advantages, even for plant species for which other options are available. A recent study on the comparative costs of maintaining a large coffee field collection with those of establishing a coffee seed cryo-collection at the Centro Agronómico Tropical de Investigación y Enseñanza (CATIE) showed that cryopreservation costs less in perpetuity per accession than conservation in field gene banks (especially if the intention is long-term storage), generating economies of scale with the costs further decreasing with an increasing number of cryopreserved accessions (Dulloo *et al.*, 2009). Nonetheless, it should be noted that cryopreservation may not prolong the storage life of germplasm indefinitely; a study revealed that while 59% of the strawberry meristems thawed were viable after 28 years of cryopreservation as compared to 56% viability after 8 weeks, only 14% of the pea meristems thawed were viable after 28 years, compared to 61% after 26 weeks of cryopreservation (Caswell and Kartha, 2009).

Cryopreservation of plant genetic resources has been reported by two countries in North America, six countries in Asia, four countries in Europe, two in Latin America and the Caribbean, one in the Near East, and none in Africa (FAO, 2010c). Among the challenges that restrict wider application of cryopreservation are the complex and time-consuming optimization of efficient protocols for new species, the differential genotypic responses to cryopreservation, and lack of knowledge regarding causal factors in cryopreservation sensitivity/tolerance. Studies are ongoing to ascertain genomic responses to cold and desiccation stresses in order to design improved cryopreservation strategies (Volk, 2010).

Cryotherapy, a modification of cryopreservation, is a novel method for pathogen eradication in which the shoot tips are briefly exposed to liquid nitrogen. It has been used mainly to eliminate viruses in banana, grapevine, potato, rasberry, sweet potato, and *Citrus* and *Prunus* species (Feng *et al.*, 2011; Wang *et al.*, 2009a).

E. *In vitro* slow growth storage

In vitro slow growth storage has been successfully applied to a range of species as well as across many genotypes within species. It markedly reduces the frequency of periodic subculturing (ranging from several months to 4 years, depending upon the species) without affecting the viability and regrowth potential of the culture. The efficiency of slow growth protocols depends upon several parameters such as the type of explants and their physiological state. The best results for establishing cultures for storage have been obtained using organized cultures such as apical meristems, axillary buds, and embryos since undifferentiated tissues such as calli are more prone to somaclonal variation (Rao, 2004; Uyoh *et al.*, 2003).

Variability between accessions in their response to culture conditions has been observed, and culture conditions need to be customized to new material. For instance, an *in vitro* core collection of African coffee germplasm with 21 diversity groups, which was conserved under slow growth, exhibited a great variability in response to the storage conditions (Dussert *et al.*, 1997). Bioversity International has developed technical guidelines for the management of *in vitro* crop germplasm collections (Reed *et al.*, 2004).

Approximately 8100 cassava accessions are conserved under slow growth conditions in 13 tissue culture banks worldwide, with 80% of these accessions in the CIAT and the International Institute of Tropical Agriculture (IITA) collections.[78] Additionally, collections of potato and sweet potato at the CIP, yam at the IITA, and banana and plantain at the INIBAP International Transit Center are stored using this method. At CIAT, stability of cassava germplasm, after *in vitro* slow growth storage for 10 years, has been ascertained through molecular analysis (Angel *et al.*, 1996). Several developing countries, including eight in Africa, have also reported having *in vitro* slow growth storage facilities (FAO, 2010c). The CGIAR System-wide Genetic Resources Program has developed a Crop Genebank Knowledge Base[79] which provides crop-specific best practices for nine crops and general procedures for different conservation methods including seed banks, field collection, *in vitro* slow growth storage, and cryopreservation.

[78]http://cropgenebank.sgrp.cgiar.org/index.php?option=com_content&view=article&id=342&Itemid=487&lang=english.

[79]http://cropgenebank.sgrp.cgiar.org/.

F. Wide crossing

Wide crossing is often employed to obtain a plant that is practically identical to the original crop with the exception of a few desirable genes contributed by the distant relative. The transfer of unwanted characteristics is circumvented through a series of backcrosses (for six or more generations), a time-consuming and laborious process. Introgression libraries, that is, marker-defined genomic regions taken from the donor parent and introgressed onto the background of elite crop lines, can help accelerate the process (Zamir, 2001).

With biotechnological approaches overcoming interspecific crossing barriers, there has been a steady increase in the utilization of CWR in plant breeding programs, with the most widespread use in the development of disease and pest resistance, followed by abiotic stress tolerance, yield increase, cytoplasmic male sterility and fertility restorers, and improved quality (Hajjar and Hodgkin, 2007; Maxted and Kell, 2009). For example, embryo rescue has been successfully used to produce hybrids between Asian rice and all other wild rice species except *O. schlechteri* (Brar, 2005). However, a future concern for wide crossing is the threat to CWR, with many species declining in distribution and abundance (Maxted and Kell, 2009). The technique of embryo rescue has also been utilized in crosses between rye and wheat resulting in the new species triticale which combines the genetic attributes of both species, that is, the adaptability to environmental conditions of rye with the yield potential and nutritional qualities of wheat (Ammar *et al.*, 2004).

In vitro embryo rescue and anther culture have been vital in the development of interspecific hybrids, including New Rice for Africa (NERICA) varieties produced by cross-breeding African rice (*Oryza glaberrima*) with Asian rice (*O. sativa*). NERICA varieties exhibit higher yields and earlier maturity than Asian rice, increased pest resistance and improved drought tolerance compared to African rice, and a higher protein content than either of the parents (FAO, 2011a; Somado *et al.*, 2008; Sonnino *et al.*, 2009). These varieties have been released in 30 African countries and have played a key role in enhanced rice harvests (Sonnino *et al.*, 2009). Approximately 300,000 hectares are estimated to be under NERICA cultivation in West, Central, and East Africa.[80]

Impact assessment studies conducted in West Africa have pointed to significantly positive impacts of NERICA adoption on rice yield in Benin and the Gambia, with higher impacts for women than for men; for example, women rice farmers had a higher additional income gain (Agboh-Noameshie *et al.*, 2007; Diagne *et al.*, 2009). Analysis in Uganda showed that NERICA has the potential to increase per capita income by USD 16 (10% of actual per capita income) and

[80]http://go.worldbank.org/OFD841GU60.

to decrease the poverty incidence with further increases in income possible when combined with effective extension services, seed delivery systems, and appropriate market development policies (Kijima *et al.*, 2008, 2011).

G. Somatic hybridization

Somatic hybridization is a technique that has been used to tap the potential of related or distant species/genera of crops and is especially useful for creating novel combinations of nuclear and/or cytoplasmic genomes. Numerous intergeneric, and intra- and interspecific somatic hybrids have been reported and genes for quality improvement, resistance against bacterial, fungal, viral and nematode diseases, as well as abiotic stress tolerance have been transferred (Liu *et al.*, 2005; Nagata and Bajaj, 2001). Several new commercial varieties of potato and oilseed rape have also been produced through this technology (Murphy, 2007).

Drawbacks of somatic hybridization include genetic instability and low fertility of the hybrids, and over the past few years, this technology has been replaced to a large extent by transgenesis. Developing protocols for somatic hybridization is often a long and cumbersome process, and poor plantlet regeneration efficiency *in vitro* is a limitation in many species. However, somatic hybridization has considerable potential for the future since, unlike transgenesis, it is more efficient for transferring polygenic traits, it does not require the same regulatory approval, and advances in tissue culture and molecular marker techniques have increased the success rate in regenerating genetically stable progeny (FAO, 2011a).

H. Micropropagation

Micropropagation is widely used for a range of vegetatively propagated subsistence crops in developing countries such as cassava and banana, commercial plantation crops such as sugarcane and oil palm, niche crops such as cardamom and vanilla, and fruit tree crops such as coconut and mango (FAO, 2011a). The most frequent application of micropropagation is for the production of virus-free plantlets through a combination of meristem culture and explant heat treatment, and subsequent mass-scale multiplication of selected plant lines or individuals (Sonnino *et al.*, 2009). However, the large-scale use of clonally propagated material can generate completely uniform crops thus further reducing the genetic diversity present in the fields, and therefore, care should be taken to include different genotypes as initial sources of plant material.

Socioeconomic impact studies, carried out in a few developing countries, indicate that the use of micropropagated material led to increased productivity and enhanced rural livelihoods. In Asia, for example, micropropagated virus-free sweet potato was adopted by 80% of the farmers in the Shandong Province of China, within a four-year period, leading to, on average, 30% increased yields and generating an estimated USD 145 million in net benefits annually (Fuglie *et al.*, 1999). In India, integrating micropropagation with disease detection led to the production of virus-free potato stocks that generated a revenue of USD 4 million over a period of 10 years (Naik and Karihaloo, 2007) as well as substantial improvement in sugarcane seedling quality and health, resulting in enhanced yields and economic returns (Jalaja *et al.*, 2008). For banana, economic analyses in India have indicated that even though the cost of production per bunch is higher for micropropagated material than for conventional planting material, the gross and net incomes from tissue culture-derived bananas are higher by 35.4% and 42.4%, respectively (Singh *et al.*, 2011a). The introduction of high-yielding and late-blight disease-resistant potato cultivars in Viet Nam boosted yields from 10 to 20 tons per hectare (Uyen *et al.*, 1996).

In Africa, commercially micropropagated crops have been reported in 12 countries (Sonnino *et al.*, 2009). In Kenya, over 500,000 farmers have planted micropropagated bananas (Wambugu, 2004) and it has been found that although micropropagated banana production is more capital intensive than traditional banana production, it offers relatively much higher financial returns (Mbogoh *et al.*, 2003). Micropropagated sweet potato varieties were taken up by 97% of the farmers in the Hwedza District of Zimbabwe leading to greater yields per hectare and net economic returns (Mutandwa, 2008; Sonnino *et al.*, 2009). Further, studies carried out in Uganda and Zimbabwe concluded that in order to maximize the socioeconomic impact of micropropagated materials, extension and training need to be complemented with service packages such as subsidized planting material, marketing, etc., and *ex ante* consideration of adoption patterns (Sonnino *et al.*, 2009).

I. Chromosome set manipulation

In plants, a rapid and cost–effective approach for inducing sterility (e.g., to produce seedless fruit) is the creation of polyploids, especially triploids. Traditionally, triploidy is induced by first producing tetraploids with colchicine treatment followed by crossing the tetraploids with the diploid counterparts, but this method is tedious and lengthy. A faster method to produce triploids is through the *in vitro* regeneration of plants from the endosperm (a naturally occurring triploid tissue) that can eventually be multiplied by micropropagation (Thomas and Chaturvedi, 2008). Using this technique, triploid varieties of several fruit crops have been created including most of the citrus fruits, acacias, kiwifruit, loquat, passionflower, and pawpaw (FAO, 2011a).

Anther culture offers excellent opportunities for large-scale production of haploids, although problems include the occurrence of albinos and induction of genetic variation. Anther culture techniques have been established for numerous economically important crop species, including vegetables (Juhasz *et al.*, 2006) and cereals such as rice, barley, and wheat (Germana, 2011). Doubled haploid (DH) plants, produced using *in vitro* anther culture (or alternatively by ovary/ovule culture) and chromosome doubling, are valuable in breeding programs since they are 100% homozygous (that is, recessive genes are readily apparent), and the need for numerous cycles of inbreeding is thus considerably reduced by shortening the time needed to select desired lines (Dunwell, 2010). However, skilled labor is required to test large populations, leading to increased costs.

Wheat cultivars derived from DHs have been released in Brazil, Canada, China, Europe, Morocco, and Tunisia (Jauhar *et al.*, 2009; Sonnino *et al.*, 2009). Anther culture techniques have also been applied to obtain DH barley (Gomez-Pando *et al.*, 2009) and rice (Pauk *et al.*, 2009) varieties better suited to the environment. In China, more than 100 new rice cultivars have been developed employing anther culture (Gueye and Ndir, 2010). The Joint FAO/IAEA division has supported and coordinated research efforts focusing on the development of reproducible DH protocols, including the publication of a manual with protocols for the production of DHs in 22 plant species (Maluszynski *et al.*, 2003).

J. Biotechnologies for disease diagnosis

The introduction of monoculture farming has resulted in a decline in genetic diversity, increasing the risk of exposure to pathogens. Diseases in crops, caused by viruses, bacteria, fungi, and nematodes, threaten food security in resource-poor countries and cause significant damage and economic losses every year (Vurro *et al.*, 2010). Early and accurate detection and diagnosis of plant pathogens is indispensable for predicting disease outbreaks and improving the precision of pesticide and/or fungicide applications as well as other control measures for disease management. Detection of pathogens is also of paramount importance in germplasm collections to enable the storage and exchange of healthy germplasm. Biotechnological diagnostic tools, such as ELISA- and PCR-based methods, for pathogen identification can bypass many shortcomings related to culture-based morphological approaches (e.g., for microbial species that are difficult to culture or to identify microscopically).

ELISA is the most widely used diagnostic technique for plant pathogens in developing countries, especially for identifying viruses (FAO, 2005). Numerous ELISA test kits that can detect diseases of root crops, fruits, cereals, and vegetables are available commercially. One of the first such ELISA kits was developed by CIP to detect the presence of *Ralstonia solanacearum*, the pathogen that causes bacterial wilt in potato.[81] Recent adaptations in the form of dipstick assays, resulting in increased convenience and portability in concert with lower costs, have further expanded the utility of these technologies in the field.

PCR is routinely employed in the developed world for diagnostic purposes (Vincelli and Tisserat, 2008). PCR-based techniques are also very effective for monitoring the emergence of novel variants of well-known pathogens, such as *Puccinia striiformis* f. sp. *tritici* that causes yellow (stripe) rust of wheat (Milus *et al.*, 2009) and *Puccinia graminis* f. sp. *tritici* that causes black (stem) rust of wheat (Visser *et al.*, 2009, 2010). Both black and yellow rust are economically damaging diseases affecting wheat production across the world, and there has been a recent escalation in the threat posed by them. The Rust SPORE web portal[82] was launched by FAO and partners in 2010 to mitigate the threat of wheat rust diseases by delivering surveillance information, monitoring pathogens, and providing access to information tools.

Real-time PCR has high sensitivity and specificity but generally requires expensive laboratory equipment. However, the cost has decreased in recent years and portable thermocyclers have been developed, for example, for the on-site diagnosis of bacterial diseases in watermelon (Schaad *et al.*, 2001) and grape (Schaad *et al.*, 2002). The Cereal Disease Laboratory (United States of America) is currently developing real-time PCR-based assays for the identification of Ug99 and other races of the wheat stem rust pathogen.[83]

The aforementioned assays are restricted in the number of pathogens that can be tested at once, while plants could be infected by several pathogens, some of which may act synergistically to cause a disease complex. Microarrays have been used for discriminating serotypes and subgroups of viruses as well as for species of fungal pathogens (Mumford *et al.*, 2006; Nicolaisen *et al.*, 2005), although existing microarray methods are complex, expensive, and consequently out of reach of many developing countries. An area where microarrays could be applied in the future is in quarantine systems. For example, the

[81] http://www.isaaa.org/resources/publications/pocketk/22/default.asp and http://www.cipotato.org/potato/pests-and-disease.

[82] http://www.fao.org/agriculture/crops/rust/stem/en/.

[83] http://www.ars.usda.gov/ug99/actionplan.pdf.

European Union (EU) funded project DiagChip aims to develop a diagnostic chip for the simultaneous detection of all the EU-listed quarantine potato pests/pathogens.[84]

Many plant disease diagnostic networks have been established to address issues related to disease diagnosis and pathogen detection, in particular, the gap in capacity between developed and developing countries (Miller *et al.*, 2009). For example, the Global Plant Clinic[85] has a network of 81 clinics in 10 countries across Africa, Asia, and Latin America (Boa, 2010). It links stakeholders like diagnostic laboratories and researchers with clinics and also runs a global diagnostic service in the United Kingdom for further testing to identify new diseases (Wilson, 2010).

The International Plant Protection Convention is a multilateral treaty that aims to increase international cooperation in plant protection by developing International Standards for Phytosanitary Measures (ISPMs).[86] Among the 34 ISPMs that have been adopted so far, one is on "diagnostic protocols for regulated pests" and provides the minimum requirements for reliable and official diagnosis of pests, including guidance on the use of morphological and molecular/biochemical diagnostic techniques.[87] The protocols also offer additional methods to be used in different circumstances such as general surveillance, diagnosis for pests found in imported consignments, and the first diagnosis of a pest in an area. In response to the need for regional harmonization, several Regional Plant Protection Organizations, such as the Comité de Sanidad Vegetal del Cono Sur,[88] the European and Mediterranean Plant Protection Organization,[89] and the North American Plant Protection Organization,[90] have developed diagnostic protocols of regulated pests in their respective regions (Clover *et al.*, 2010).

K. Molecular marker-assisted selection

MAS has the potential to increase genetic gain by permitting selection at an earlier stage of development and/or by reducing the generation interval or the number of generations needed prior to releasing a new variety. It is especially advantageous when phenotypic recording is destructive or pyramiding of genes is

[84]http://www.diagchip.co.uk/.

[85]A consortium of CABI Bioscience, Rothamsted Research, and Central Science Laboratory, United Kingdom.

[86]https://www.ippc.int/.

[87]ISPM No. 27—https://www.ippc.int/file_uploaded/1155903234858_ISPM27_2006_E.pdf.

[88]http://www.cosave.org/estandares.php?ver=3.

[89]http://archives.eppo.org/EPPOStandards/diagnostics.htm.

[90]http://www.nappo.org/Standards/Std-e.html.

desired. It is particularly useful for horticultural crops, since most of them are highly heterozygous which makes phenotypic selection difficult (Ibitoye and Akin-Idowu, 2010).

MAS has been used in plant breeding for developing new hybrids and varieties of both annual crops such as cereals and legumes and perennial crops such as fruit trees, tea, and coffee (Collard and Mackill, 2008; FAO, 2007b, 2011a). For a commercial maize breeding program in Europe and North America, MAS has been shown to increase the mean performance of progeny for multiple traits such as grain yield and moisture content compared to conventional breeding (Eathington *et al.*, 2007).

Sorghum varieties have been developed that are resistant to striga, a weed that infests nearly 100 million hectares of field crops in Africa annually. Marker-assisted backcrossing was used to introgress target genes into locally adapted landraces and improved sorghum lines, with broad adaptation and high yield, to produce agronomically superior cultivars that are being grown in several African countries (Ejeta, 2007). Integrated striga management, combining the use of striga-resistant cultivars with soil fertility management and moisture conservation, has further increased sorghum productivity (Ejeta and Gressel, 2007).

Quality protein maize (QPM) contains approximately twice as much usable protein as regular maize grown in the tropics and is of particular use in developing countries to fight malnutrition (Krivanek *et al.*, 2007). MAS has been utilized to improve the efficiency of selection to develop QPM hybrids (Babu *et al.*, 2005; Danson *et al.*, 2006), and commercial cultivars have been released in India (Gupta *et al.*, 2009). Other successful examples of hybrids and varieties released using MAS include a pearl millet hybrid resistant to downy mildew in India (Dar *et al.*, 2006), drought-resistant soybean in the United States of America,[91] disease-resistant barley varieties in Australia (Eglinton *et al.*, 2006), rice varieties with resistance to bacterial blight in China, India, Indonesia, and the Philippines (Vogel, 2009), with low amylase in the United States of America (Dwivedi *et al.*, 2007), with submergence tolerance in Bangladesh, India, and the Philippines,[92] and with drought tolerance in India (Vogel, 2009).

In spite of its high potential, MAS is still applied in relatively few breeding programs in developing countries. This is because an effective MAS strategy requires adequate laboratory capacity and data management, trained personnel, and operational resources (Ribaut *et al.*, 2010). Another factor has

[91]http://www.wisconsinagconnection.com/story-national.php?Id=808&yr=2009.
[92]http://www.scidev.net/en/news/india-takes-to-new-flood-tolerant-rice.html.

been the scarcity of genetic and genomic resources for orphan crops that play an important role in developing countries, although international collaborations and initiatives are now addressing these issues (Varshney *et al.*, 2009b). The CGIAR GCP has also developed a Molecular Marker Toolkit[93] that allows access to current effectively used markers for application in MAS for 12 food security crops (Van Damme *et al.*, 2011).

The relative costs of applying MAS are higher than conventional approaches. Studies comparing MAS with conventional selection suggest that the cost–effectiveness of both methods depends on the particular circumstances of the specific application and that the optimal choice of the breeding application should be based on a case-by-case analysis (Dreher *et al.*, 2003; Morris *et al.*, 2003; William *et al.*, 2007). However, MAS is becoming progressively cheaper and more cost–effective.

A recent *ex ante* impact analysis from the CGIAR, taking into account crop yields, farmer adoption rates, market prices, cultivated land area, breeding and dissemination times, input prices, and costs of marker development, concluded that MAS in rice and cassava will result in significant incremental benefits over conventional breeding.[94] For example, MAS for tolerance to salinity and phosphorus deficiency in rice is expected to save 3–6 years with projected economic benefits, over 25 years, of USD 50–900 million, depending upon the country, stress, and time lag (Alpuerto *et al.*, 2009). For cassava, MAS for resistance to pests and post-harvest physiological deterioration is estimated to save 4 years with net benefits over 25 years in the range of USD 34–800 million contingent on various assumptions (Rudi *et al.*, 2010).

The incorporation of genomic information is further underpinning crop improvement, and MAS is evolving into GS in some developed countries (Heffner *et al.*, 2009; Jannink *et al.*, 2010). Simulation studies have shown the accuracy of estimating the breeding value from GS to be comparable to phenotypic selection, without entailing the time and expense of field evaluation (Zhong *et al.*, 2009). A recent study indicated that even at low accuracy, annual gain from GS exceeded that of MAS about threefold in a high-investment maize breeding program and twofold in a low-investment winter wheat breeding program (Heffner *et al.*, 2010). GS, thus, has the potential to lower costs and increase rates of genetic gain, leading to more effective selection strategies (Heffner *et al.*, 2011). GS models for adult pathogen resistance to wheat stem rust are currently being developed as a promising strategy in the fight against the disease (Rutkoski *et al.*, 2011).

[93]http://s2.generationcp.org/gcp-tmm/web/.
[94]http://www.generationcp.org/sp5_impact/exante-norton-conclusions.

L. Mutagenesis

Induced mutagenesis has played an important role in the development of superior crop varieties by generating greater genetic diversity in existing varieties. It is particularly important for the genetic improvement of vegetatively propagated crops where cross-breeding is not possible or is time-consuming (Mba *et al.*, 2009). Induced mutagenesis offers the possibility of introducing desired attributes that have either been lost during evolution or are not present in nature. Another major advantage is the ability to isolate mutants with multiple traits as well as mutant alleles with varying degrees of trait modification. The resultant mutant varieties can be readily commercialized without the regulatory requirements applied to transgenic crops, though the limitation remains that mutagenesis can only be used to manipulate already existing genes, usually by suppressing/deleting their function (Parry *et al.*, 2009).

Somaclonal mutagenesis refers to the epigenetic or genetic alterations induced during *in vitro* culture. Typical DNA changes include chromosome number changes, chromosomal rearrangements (e.g., translocations, deletions, insertions, and duplications), point mutations, and gene methylation or demethylation. Somaclonal variation is generally considered an undesirable by-product of the stress imposed by tissue culture, but when carefully controlled can provide novel genetic variations to plant breeders. An added benefit is that it can be used to generate variation in vegetatively propagated plants, which are usually less amenable to mutation breeding. This technology has been applied to manipulate various traits for enhancing crop productivity such as disease and pest resistance, drought and salt tolerance, and improved nutritional quality, and cultivars have been released in a few crops such as rice, wheat, maize, potato, tomato, and sweet potato (Jain, 2001).

As indicated earlier, the Joint FAO/IAEA Division has made irradiation technology more widely available to developing countries through extensive research and development as well as training and capacity development activities (Lagoda, 2009). Almost 3000 improved crop varieties in about 170 species have been developed through induced mutation and released in an estimated 100 countries generating economic benefits for farmers (FAO/IAEA, 2008). For example, three improved varieties of rice produced a total net profit of USD 348 million for farmers in Viet Nam in 2007 alone, while in Peru, the introduction of nine superior barley varieties has resulted in 50% increased harvests translating to roughly USD 9 million a year (IAEA, 2008).

The most commonly used chemical mutagens are alkylating agents such as ethylmethane sulfonate and *N*-methyl-*N*-nitrosourea that induce point mutations in DNA. Since point mutations are less detrimental than large chromosomal rearrangements, this method has a higher frequency of achieving a saturated mutant population (Gilchrist and Haughn, 2010). However, point mutations are often recessive, and therefore, the second or later generations of mutagenized tissues must be screened to identify homozygous recessive mutations. Chemical mutation-derived varieties have been obtained and commercially released for numerous staple species including rice, wheat, maize, soybean, and barley.[95]

A recent approach that combines classical mutagenesis with high-throughput identification of mutations is TILLING (targeting induced local lesions IN genomes). DNA from a collection of mutagenized plants is pooled, subjected to PCR amplification, and screened for mutations by detecting mismatches in duplexes with nonmutagenized DNA sequences (McCallum *et al.*, 2000). TILLING is particularly advantageous as mutations can be detected in pools of small plantlets without the need to screen adult plants for an observable phenotype. It is also amenable to automation, making it especially conducive for crop species that have large and complex polyploid genomes. However, TILLING requires prior DNA sequence information and is a labor-intensive technique. Further, the availability of a mutagenized population is a prerequisite for TILLING, and the development of such populations is expensive and time-consuming for many species.

TILLING platforms and associated large mutagenized populations, valuable resources for screening for traits of interest, have been created for several crops including rice[96]; maize[97]; durum wheat[98]; barley[99]; and tomato, rapeseed, and pea.[100] The Joint FAO/IAEA Division has developed lower-cost assays and kits for mutation discovery that are affordable for laboratories in developing countries.

TILLING with vegetatively propagated crops, which are less genetically tractable, is still in its infancy. Among the challenges involved in developing a suitable mutagenized population are the choice of mutagen as well as the tissue to be mutagenized. The Joint FAO/IAEA Division is currently establishing TILLING platforms for banana, cassava, and yam, for improving food security in developing countries.[101]

A variation on TILLING is EcoTILLING, which looks for polymorphisms in natural populations (Comai *et al.*, 2004). It is highly informative for species that are not amenable to mutagenesis and has been applied for allele

[95]http://mvgs.iaea.org/.
[96]http://tilling.ucdavis.edu/.
[97]http://genome.purdue.edu/maizetilling/.
[98]http://www.rothamsted.bbsrc.ac.uk/cpi/optiwheat/indexcontent.html.
[99]http://www.gabi-till.de/project/ipk/barley.html and http://www.distagenomics.unibo.it/TILLMore/.
[100]http://www.versailles.inra.fr/urgv/tilling.htm.
[101]http://www-naweb.iaea.org/nafa/pbg/public/pbg-nl-25.pdf.

mining for variation in disease resistance (Barkley and Wang, 2008). It has also been used to detect genetic diversity in inbred lines, cultivars, and accessions of agronomically important crop plants (Weil, 2009).

M. Transgenesis

Genetically modified crops were first grown commercially in the mid-1990s. In 2010, transgenic crops are estimated to have been cultivated on 148 million hectares in 18 developing and 11 industrialized countries, with 45% of the global total being cultivated in the United States of America (James, 2010). Seventeen countries, including 13 developing countries, planted over 50,000 hectares each. The four main transgenic crops grown were soybean, maize, cotton, and canola, with herbicide tolerant soybean being the principal crop. Further details on the application of transgenic crops can be found in FAO (2011a).

IV. CURRENT STATUS OF BIOTECHNOLOGIES FOR THE MANAGEMENT OF FOREST GENETIC RESOURCES

Forests provide a vast array of economic, environmental, and social products and services. Thirty percent of the world's forests are primarily used for the production of wood and nonwood forest products, while 8% are designated for the protection of soil and water resources (FAO, 2010b). Forests also play a very significant role in carbon sequestration, storing an estimated 289 gigatons of carbon in their biomass. The world's total forest area is just over 4 billion hectares, that is, 31% of the total land area, with primary forests accounting for 36% and planted forests 7% of the total forest area (FAO, 2010b). Approximately 10 million people are employed in forest management and conservation, but millions more rely on forests to a high degree for their livelihoods.

Forests possess much of the world's biodiversity, but this diversity is threatened by a high rate of deforestation due to an expansion in global agricultural and industrial needs. Other threats include diseases, pests, and weeds, many of which are introduced from other regions. Forest trees also have certain characteristics that differentiate them from other agricultural sectors, such as crops and livestock, for example, their highly heterozygous nature, long generation intervals, vulnerability to inbreeding depression, narrow regional adaptation, and the fact that the majority of the species are undomesticated (FAO, 2011b)—thus generating unique challenges and opportunities for biotechnology applications. As elaborated in this section, biotechnological approaches have advanced considerably in the past decade and have contributed to creating more efficient and effective characterization, conservation, and utilization strategies for forest genetic resources. However, in spite of their importance in natural ecosystems, genetic resources of noncommercial species, encompassing also lower plants, have received less attention, and therefore, there is a pressing need to focus on their sustainable management, including through biotechnological approaches.

A. Molecular markers

Molecular markers can be used to study the level, structure, and origin of genetic variation in both naturally regenerated tropical forests and planted forests. Molecular marker studies, thus far, have suggested that natural populations of most tropical tree species contain higher levels of variation relative to other plants (Kindt *et al.*, 2009). However, the extent of genetic diversity in tropical forests is still largely unknown, and many economically important species are yet to be identified. Molecular markers have helped to differentiate and inventory species, but a thorough cataloging will require good linkages with field botanists (Dick and Kress, 2009). For tropical trees, a practical protocol guide on molecular marker methods (Muchugi *et al.*, 2008a) as well as an accompanying guide on the effective handling and analysis of the datasets generated (Kindt *et al.*, 2009) have been produced by the World Agroforestry Centre.

To develop effective conservation strategies, accurate taxonomic identification together with the ability to discriminate hybrids from pure species and to estimate the degree of introgression is essential (Wang and Szmidt, 2001). The phylogenetic relationships of tree species that are difficult to distinguish on the basis of morphological characteristics alone can be resolved in conjunction with the use of molecular markers, for example, between species of *Warburgia* (Muchugi *et al.*, 2008b), *Populus* (Cervera *et al.*, 2005), and *Quercus* (Zeng *et al.*, 2010).

Molecular markers have facilitated the identification of natural hybrids across species thus leading to a better understanding of introgression, for instance, between *Fraxinus excelsior* and *F. angustifolia* (Fernández-Manjarrés *et al.*, 2006), *Populus alba* and *P. tremula* (Lexer *et al.*, 2005), *Pinus echinata* and *P. taeda* (Xu *et al.*, 2008), and *Quercus suber* and *Q. ilex* (Burgarella *et al.*, 2009). Furthermore, natural hybrid zones are valuable for investigating evolutionary processes of speciation (Tovar-Sánchez and Oyama, 2004) as well as for identifying QTLs for adaptive genetic variation in forest trees (Lexer *et al.*, 2004).

Determination of the origin of forest reproductive material for planted forests as well as traded wood and wood products is crucial. Molecular markers have been applied to improve traceability through reliable identification of species to control both trade with protected trees and illegal logging (Finkeldey *et al.*, 2010). Recently, a standard two locus DNA barcode has been proposed for species discrimination of land plants, including forest trees (CBOL Plant Working Group, 2009).

Molecular markers have offered insights into the domestication of forest trees, such as the origin of olive from the oleaster (Besnard *et al.*, 2001; Breton *et al.*, 2006). Additionally, based on nuclear and cytoplasmic markers, nine domestication events have been proposed for olive cultivars (Breton *et al.*, 2009), while AFLP data have suggested two distinct geographic origins of cultivated *Spondias purpurea* trees in Mesoamerica (Miller and Schaal, 2006). Molecular marker studies have also shed light on the evolutionary history of tree species, for instance, the postglacial routes of colonization of *Populus nigra* occurred from two main refugia in Italy and/or the Balkans and Spain (Cottrell *et al.*, 2005).

Since most forest trees are outcrossing, they do not generally show evidence of strong genetic differentiation among populations and the highest genetic diversity is found within populations. Molecular markers have been used to measure genetic variation in tropical trees, for example, within and between populations of *Calycophyllum spruceanum* in the Peruvian Amazon Basin (Russell *et al.*, 1999), *Sesbania sesban* in sub-Saharan Africa (Jamnadass *et al.*, 2005), *Tectona grandis* in India (Narayanan *et al.*, 2007) as well as diverse geographical regions in Indonesia and Thailand (Shrestha *et al.*, 2005), and *Guaiacum sanctum* in Costa Rica (Fuchs and Hamrick, 2010). Similar analyses on molecular genetic diversity have been carried out in temperate tree species, for example, within and among populations of *Fagus grandifolia* in Mexico (Rowden *et al.*, 2004), *Robinia pseudoacacia* in China (Huo *et al.*, 2009), and *Sorbus torminalis* in Europe (Rasmussen and Kollmann, 2008). Altitudinal variation has also been observed within populations of tree species, indicating that both vertical and horizontal patterns of genetic diversity must be considered while designing conservation strategies (Ohsawa and Ide, 2008).

Molecular markers have been utilized to evaluate the efficiency of agroforestry systems for conservation of forest genetic resources by comparing the genetic variation across natural forest and proximate planted farm stands. RAPD studies of the timber tree Meru oak in central Kenya showed little differentiation between unmanaged and managed stands (Lengkeek *et al.*, 2006). Another study assessing the genetic diversity between planted and natural stands of *Inga edulis* from five sites in the Peruvian Amazon demonstrated lower allelic variation in planted stands, even though on-farm stands contained on average 80% of the allelic diversity of natural stands (Hollingsworth *et al.*, 2005). An explanation for the difference between the two studies could be that while all on-farm *I. edulis* was of planted origin, the oak trees in Kenya may have been planted or naturally regenerated.

Exploration of the amount and distribution of genetic variation in clonally propagated domesticated stands and sexually reproducing wild populations of *S. purpurea* revealed that levels of genetic variation within cultivated stands were significantly lower than in wild populations (Miller and Schaal, 2006). Moreover, within the cultivated populations, trees in orchards harbored less genetic variability than trees in backyard gardens and living fences.

Knowledge of mating systems and gene flow is important for understanding genetic drift, natural selection, and population divergence and for designing conservation strategies that maximize connectivity of populations in fragmented forests while minimizing unwanted gene flow. For example, microsatellite data evaluating the mating system and pollen gene flow in oak in northern Thailand showed high outcrossing rates, high levels of gene flow from outside populations, and heterogeneity in the pollen composition received by individual trees suggesting that losses of genetic diversity of the species could be prevented at the study site (Pakkad *et al.*, 2008). Population genetic diversity in disturbed and undisturbed teak forests within the natural range of the species, the mating system, and contemporary gene flow has also been studied using molecular markers (Volkaert *et al.*, 2008).

Maternally and paternally inherited markers make it possible to distinguish the separate contributions from pollen and seed in gene flow studies (Dick *et al.*, 2008; Hamza, 2010; Jones *et al.*, 2006; Sork and Smouse, 2006). Molecular information has demonstrated that gene flow through pollen dispersal is significantly higher (20 to nearly 200 times) than gene flow through seeds, at least among wind-pollinated species (Savolainen *et al.*, 2007) and tree species that produce large, immobile seeds (Dow and Ashley, 1998). Molecular markers have been used to study pollen-mediated gene flow in populations of both wind- and animal-pollinated trees (Burczyk *et al.*, 2004), with long distance pollen dispersal (over 5–10 km) not uncommon (Petit and Hampe, 2006), which could have significant implications for the conservation of trees in fragmented stands (Kamm *et al.*, 2009; White *et al.*, 2002).

Gene flow from planted to natural stands, that is, anthropogenic hybridization as a result of human activity, can have a profound effect on the diversity and adaptability of wild populations. Most poplar plantations in China, Europe, and North America represent a very narrow genetic base (since they are clonally propagated) and could lower the effective population size and alter the evolutionary potential of native poplar populations. Molecular markers have provided evidence for gene flow between cultivated poplars and native black poplar trees, with the frequency of hybridization depending upon the size of the native population compared to the cultivated plantations (Broeck *et al.*, 2004, 2005). Extensive hybridization has also been observed between the native North American butternut and the introduced Japanese walnut tree (Hoban *et al.*, 2009).

Molecular markers can be used to manage clonally propagated domesticated stands by aiding the selection and identification of clones (Aravanopoulos, 2010; Guan *et al.*, 2010; Hiraoka *et al.*, 2009; Toral Ibañez *et al.*, 2009) and verifying the genetic stability of propagated material (Chandrika and Rai, 2009; Gangopadhyay *et al.*, 2003; Huang *et al.*, 2009b; Lopes *et al.*, 2006). The application of molecular markers for clonal identification and ensuring the genetic fidelity of the mass propagated clones has led to the production of superior teak clones in Malaysia (Goh *et al.*, 2007). Molecular markers are routinely used for the correct identification of clones in commercial *Eucalyptus* breeding and production forestry in Australia, Brazil, Chile, Portugal, Spain, and South Africa (Grattapaglia, 2008a).

QTL mapping is complicated in forest trees due to growth under conditions of great environmental heterogeneity, long generation intervals, large genome sizes, small segregating populations, and lack of multigenerational pedigrees. Nevertheless, QTLs for a variety of traits, such as disease resistance, drought and cold tolerance, wood quality, and bud phenology, have been mapped in numerous tree species (FAO, 2007b), with growth-related traits being the main targets in tree breeding and plantation forestry (Grattapaglia *et al.*, 2009). QTL validation is necessary to identify stable QTLs that are potentially more useful for MAS in forest trees and is carried out based on repeated QTL detection in additional individuals from the same family or among populations and multiple growing seasons (Brown *et al.*, 2003; Devey *et al.*, 2004). Alternatively, QTLs can be validated by comparative mapping between species or genera (Casasoli *et al.*, 2006; Chagné *et al.*, 2003).

B. "Omics"

Thus far, *Populus trichocarpa* (Tuskan *et al.*, 2006) and *Eucalyptus grandis*[102] are the only forest trees for which the genome sequence has been completed. Currently, genome sequencing of other *Populus* species and Fagaceae and Pinaceae species is also underway (Neale and Kremer, 2011). Based on available DNA sequences, it has been suggested that trees have higher rates of genome-wide recombination (correlated with higher levels of genetic diversity) than short-lived herbs and shrubs, with the exception of conifers which exhibit lower recombination rates than angiosperms (Jaramillo-Correa *et al.*, 2010).

SNP frequency in tree species that have been surveyed is high, approximately 1 SNP/100 base pairs, and their discovery has been facilitated by sequencing expressed sequence tags (ESTs) and candidate genes (Külheim *et al.*, 2009; Novaes *et al.*, 2008; Parchman *et al.*, 2010; Ueno *et al.*, 2010). High-throughput SNP genotyping coupled with the candidate gene approach has been used for association mapping with phenotypes of interest and aided in the dissection of complex traits such as wood quality, drought or cold tolerance, and disease resistance (Dillon *et al.*, 2010; Eckert *et al.*, 2009a,b). DArT arrays (see Section III.B), with over 8000 markers, have been developed for population and phylogenetic studies within and between species of *Eucalyptus* (Sansaloni *et al.*, 2010; Steane *et al.*, 2011).

[102]http://greenbio.checkbiotech.org/news/eucalyptus_tree_genome_deciphered.

Comparative mapping studies have uncovered extensive synteny and colinearity in conifers (Krutovsky *et al.*, 2004; Pelgas *et al.*, 2006), together with small chromosomal disruptions (Shepherd and Williams, 2008) leading to deeper understanding of speciation. The comparative resequencing project[103] is developing resources for Pinaceae comparative genomics. Comparative genomics tools have also become available in *Populus* (Douglas and DiFazio, 2010; Neale and Ingvarsson, 2008) and *Eucalyptus* (Külheim *et al.*, 2009; Paiva *et al.*, 2011).

Transcriptome profiling is particularly challenging in tree species due to their large genome sizes and lack of reference sequences. In spite of this, it has been utilized to study growth (Grönlund *et al.*, 2009; Park *et al.*, 2008), adaptation to biotic (Azaiez *et al.*, 2009; Heller *et al.*, 2008) and abiotic (Holliday *et al.*, 2008; Kreuzwieser *et al.*, 2009) stress, and wood formation (Paiva *et al.*, 2008; Wang *et al.*, 2009b). Comparative transcriptomics is being employed to study the molecular basis of complex traits such as drought tolerance (Cohen *et al.*, 2010) and fungal resistance (Barakat *et al.*, 2009). Additionally, transcriptomic data, together with linkage mapping, are being used for the identification of candidate genes (Kirst *et al.*, 2004; Sederoff *et al.*, 2009). Proteomics research in forest trees is still limited and restricted to a few genera (Abril *et al.*, 2011).

C. Bioinformatics

The principal repository of forest tree genomic data is maintained by the Dendrome Project.[104] The associated TreeGenes database[105] provides information on EST sequences, SNPs, genetic maps, molecular markers, phenotypes, and QTLs, as well as tools for their analysis. It focuses primarily on conifers, but resources for *Populus* and *Eucalyptus* are also being integrated (Wegrzyn *et al.*, 2008). *Populus*-specific databases include the RIKEN *Populus* database[106] that contains information on 10 *Populus* species, the database of poplar transcription factors[107] which currently contains 2576 putative transcription factors gene models distributed in 64 families, and the *Populus* Genome Integrative Explorer[108] with expression tools as well as browser tools for synteny and QTLs (Yang *et al.*, 2009a). *Eucalyptus*-specific resources include the International Eucalyptus Genome Network[109] and the Brazilian Network of Eucalyptus Genome Research (Grattapaglia, 2008b).

[103]http://dendrome.ucdavis.edu/crsp.
[104]http://dendrome.ucdavis.edu.
[105]http://dendrome.ucdavis.edu/treegenes/.
[106]http://rpop.psc.riken.jp.
[107]http://dptf.cbi.pku.edu.cn/.
[108]http://www.popgenie.org/.
[109]http://web.up.ac.za/eucagen/.

D. Cryopreservation

The majority of recalcitrant seeds have been identified in trees and shrubs, with approximately 47% of the species from evergreen rain forests having seeds that are desiccation sensitive (Tweddle *et al.*, 2003). Thus, cryopreservation is especially important for the long-term conservation of forest germplasm. The role of cryopreservation is further highlighted in situations where it may be complicated to find natural stands that are diverse enough for *in situ* conservation. A disadvantage of this technique is the overall difficulty associated with the regeneration of whole trees.

Most forest trees are still undomesticated and cryopreservation protocols have thus been developed and/or optimized for only a limited number of genotypes. Within these, various tissues of softwood and hardwood species have been successfully cryopreserved including embryos, embryogenic cultures, seeds, pollen, and shoot tips (Panis and Lambardi, 2006). Cryopreservation (predominantly of shoot tips) is being increasingly applied for hardwood trees such as *Populus*, *Robinia*, *Betula*, *Quercus*, *Fraxinus*, *Morus*, and *Eucalyptus* (Haggman *et al.*, 2008).

In softwood tree species, cryopreservation has been reported for more than 10,000 genotypes of over 23 conifer species and their hybrids (Tsai and Hubscher, 2004). In conifer clonal forestry, it is used as a suitable and efficient means for the storage of embryogenic cultures awaiting field testing results. Clonal varieties can then be developed by thawing and propagating the desired cryopreserved embryogenic tissue clones that are superior in the field tests (Park, 2002; Sharma, 2005).

So far, there is little evidence of genetic alterations in forest trees caused by cryopreservation. Embryo recovery levels for cryopreserved oak embryogenic lines ranged from 57% to 92%, with no genetic instability observed in the regenerated plants (Sanchez *et al.*, 2008). Similarly, the genetic fidelity of silver birch meristems and mulberry axillary winter-dormant buds was maintained subsequent to cryopreservation (Atmakuri *et al.*, 2009; Ryynanen and Aronen, 2005). The viability of cryopreserved material has also been determined. Dormant European ash seeds, cryopreserved for two years following desiccation to a safe water content, did not exhibit decreased germination after thawing relative to a 2-year seed storage at −3 °C (Chmielarz, 2009). Likewise cryopreserved pecan pollen, stored for 1–13 years, did not demonstrate reduced viabilty compared to fresh pollen (Sparks and Yates, 2002).

Implementation of cryopreservation in developing countries has been restricted due to economic constraints. However, cryopreservation of dormant elm buds has been shown to be economically competitive to field clonal archives, with a twofold cost saving in favor of the cryobank (Harvengt *et al.*, 2004). Cryopreservation of forest trees has been initiated in some developing countries, for example, for pollen of tree species at the National Bureau of Plant Genetic

Resources, India, and for species such as *Dipterocarpus*, *Bambusa*, and *Dendrocalamus* in Indonesia (Jalonen *et al.*, 2009). Cryogenic repositories for forest tree species include 420 accessions of mulberry at the National Institute of Agrobiological Resources, Japan and 440 elm accessions at Association Forêt-Cellulose, France (Engelmann, 2011).

E. *In vitro* slow growth storage

In vitro slow growth storage techniques require establishment of specific protocols depending on the type of explant and species under consideration. Another issue to be considered with tropical tree species is the presence of endophytes[110] that can cause difficulties for the establishment of sterile cultures (Muralidharan and Kallarackal, 2005).

Successful protocols have been developed for over 30 woody species, including *Pinus radiata*, *Alnus glutinosa*, and species of *Eucalyptus* and *Populus* (FAO, 1994). *In vitro* cultures of *Melia azedarach* apical meristem tips can be maintained for 1 year without subculture or addition of fresh medium (Scocchi and Mroginski, 2004), of *E. grandis* shoot for up to 10 months (Watt *et al.*, 2000), 60 months for *Populus* species (Hausman *et al.*, 1994), and 6 months for *Cedrus* species (Renau-Morata *et al.*, 2006). In Spain, an *in vitro* collection, with 32 high-quality clones, has been established with selected European aspen (Martin *et al.*, 2007).

F. Micropropagation

Clonal propagation of commercially important tree species is essential in production forestry, for both coniferous and hardwood species, to provide clones of mature and elite genotypes. Additional advantageous aspects of micropropagated trees, compared to trees produced from seedlings, include more uniformity in height and trunk girth, increased biomass production, and reduced bark fissuring (Muralidharan and Kallarackal, 2005).

Micropropagation techniques have been applied to over 80 genera of forest trees, with five genera, that is, *Pinus*, *Picea*, *Eucalyptus*, *Acacia*, and *Quercus*, accounting for 50% of the documented activities (FAO, 2004). Micropropagation activities have been reported to be most numerous in Asia (38%), followed by Europe (33%), North America (16%), South America (7%), Africa (3%), Oceania (2%), and the Near East (1%) (FAO, 2004).

[110]An endosymbiont, often a bacterium or fungus, that lives within a plant for at least part of its life without causing apparent disease.

Several endogenous and exogenous factors influence *in vitro* growth and the eventual success of micropropagation. In addition to the seasonal effect, type, and age of the explants, genotype is a crucial factor in determining the responsiveness of the material to micropropagation (Durkovic and Misalova, 2008; Mashkina *et al.*, 2010; Yasodha *et al.*, 2004). *In vitro* multiplication of desired genotypes can be achieved via axillary budding, adventitious budding, or somatic embryogenesis.

Axillary budding refers to the propagation of plants through shoot development from cultured axillary buds.[111] It is the most successful clonal technique for angiosperms and produces the most true-to-type plantlets. Moreover, multiplication rates per subculture cycle can be higher than in adventitious budding. Protocols using this method have been developed for several species (Pijut *et al.*, 2011) including *P. tremula* (Peternel *et al.*, 2009), *T. grandis* (Shirin *et al.*, 2005), *Dalbergia sissoo* (Thirunavoukkarasu *et al.*, 2010) as well as species of *Eucalyptus* (Arya *et al.*, 2009; Glocke *et al.*, 2006), *Acer* (Durkovic and Misalova, 2008), and *Quercus* (Vieitez *et al.*, 2009).

Adventitious budding refers to the induction of adventitious[112] buds on nonmeristematic tissue and is the preferred method for micropropagation of conifers. Induction rates can be quite high, but it is more prone to somaclonal variation. Species of *Pinus* (Alonso *et al.*, 2006; Álvarez *et al.*, 2009), *Prunus*, *Ulmus*, and *Fraxinus* (Durkovic and Misalova, 2008), among others, can be propagated by this technique.

Somatic embryogenesis (SE) is the process of differentiation of somatic embryos from vegetative cells, through the application of exogenous growth regulators to juvenile tissue. SE systems have been developed for both conifer species (Nehra *et al.*, 2005) and temperate wood species (Pijut *et al.*, 2007, 2011), with conifer species usually being more intractable (Bonga *et al.*, 2010). A key advantage of SE is that the embryogenic tissue can be cryopreserved without loss of viability or genetic integrity, while corresponding trees are field-tested (Park, 2002). It has the largest potential multiplication rate and is amenable to handling in automated bioreactors (FAO, 2011b). However, SE is expensive and commercial application of this technique is still limited.

In Malaysia, collaboration between the Sabah Foundation Group and Centre de Coopération Internationale en Recherche Agronomique pour le Developpement (CIRAD, France) led to the development of superior teak clones, with respect to intrinsic wood qualities (Goh *et al.*, 2007). These clones are also being exported, and preliminary data from trials in Australia, Brazil, and the United Republic of Tanzania have indicated that they outperform

[111] A bud found at the axil of a leaf.

[112] A structure arising at sites other than the usual ones, for example, shoots from roots or leaves and embryos from any cell other than a zygote.

clones from other sources, displaying a 30% increase in yield (Goh *et al.*, 2010). *Eucalyptus* hybrid clones constitute a considerable portion of existing commercial plantations, particularly in South America. Clonal forestry of selected *E. grandis* hybrid clones has been shown to reduce wood-specific consumption[113] by 20% while second generation clones derived from hybridization with *E. globulus* have led to a further reduction of 20% (Grattapaglia and Kirst, 2008).

A criticism of clonal forestry has been that it can reduce genetic diversity and make clonal plantations vulnerable to unexpected outbreaks of diseases and pests. Deploying more clonal lines could lower the risk, but also reduce genetic gain. Therefore, an appropriate balance between genetic gain and diversity is essential and it has been proposed that 10–30 clones mixed in a plantation should be sufficient to achieve this balance, in concert with a suitable configuration of deployed clones (Park and Klimaszewska, 2003).

G. Chromosome set manipulation

Due to their long regeneration time and strong inbreeding depression, forest tree species especially stand to benefit from the production of DH plants. In addition to shortening the breeding period, production of DH trees is beneficial for the isolation of recessive traits at sporophytic level. Induced haploid production through anther culture has been reported for about 32 woody species, including *Populus* and *Quercus*, but the success rate has been marginal and efficient anther culture systems are still limited (Andersen, 2005). Forest trees have been shown to be extremely intractable in anther cultures and a major impediment has been the conversion of calli and embryos into plantlets (Srivastava and Chaturvedi, 2008).

Triploids are of economic value in forest trees since they have more vigorous vegetative growth than the corresponding diploids. Triploid *Acacia nilotica* (Garg *et al.*, 1996) and mulberry trees (Thomas *et al.*, 2000) have been produced by endosperm culture, although this technique has been mostly unutilized in forest trees.

H. Molecular marker-assisted selection

Most tree breeding programs rely on recurrent selection, that is, implementing cycles of selection with intermating, generation after generation. The goal is to improve the overall performance of the population while maintaining genetic diversity, rather than to develop outstanding varieties for immediate use.

[113]The amount of wood in cubic meters necessary to produce 1 ton of pulp.

Thus, advantages of using MAS in tree breeding include reduction of generation time, decreased field-testing costs, and increased efficiency of selection for low-heritability traits (Neale and Kremer, 2011).

However, MAS is yet to realize its potential due to a number of reasons, such as the high heterogeneity of breeding populations, the lack of simply inherited traits that could be easily targeted, the difficulty in developing inbred lines to better understand the genetic basis of quantitative traits, and the limited number of scientists working in this area (FAO, 2007b; Grattapaglia and Kirst, 2008). GS has also been evaluated for tree breeding through simulation studies, and initial results have been promising but are contingent upon requirements of effective population size and genotyping density being met (Denis and Bouvet, 2011; Grattapaglia and Resende, 2011).

A recent innovative strategy in tree breeding has been "breeding without breeding" (BWB), that allows the capture of 75–85% of the genetic response to selection achieved through conventional breeding programs, but without performing any controlled crosses or experimental field testing, which are costly and time-consuming (El-Kassaby and Lstibůrek, 2009; El-Kassaby *et al.*, 2011). BWB combines phenotypic preselection of superior individuals with molecular markers for parentage analysis and pedigree reconstruction to identify elite genotypes retrospectively for establishing seed orchards. It is thus an effective and economic approach that seems to be a viable option for developing countries, especially for tree species that do not have advanced breeding programs in place (Wang *et al.*, 2010b).

I. Mutagenesis

Mutation breeding is complicated in forest tree species because recessive mutations are masked in heterozygous plants and it is difficult to obtain homozygous lines. As such, efforts have focused on the creation of dominant mutations, primarily in *Populus* species and hybrids, by activation tagging and enhancer[114] and gene traps (Busov *et al.*, 2005). Activation tagging involves insertion of strong enhancers via *Agrobacterium*-mediated transformation,[115] followed by screening of the resulting phenotypes in primary transformants and identification of candidate gene(s). Two activation-tagged populations have been created in

[114]A eukaryotic DNA sequence which increases the transcription of a gene. Located up to several kilobase pairs, usually (but not exclusively) upstream of the gene in question. In some cases can activate transcription of a gene with no (known) promoter.

[115]The process of DNA transfer from *Agrobacterium tumefaciens* to plants, that occurs naturally during crown gall disease, and can be used as a method of transformation.

Populus—a population of 627 independent lines that have undergone 2 years of field testing for mutant identification (Busov *et al.*, 2011) and a population of 1800 independent lines (Harrison *et al.*, 2007).

Alternative techniques to produce dominant phenotypes include gene and enhancer trapping. Gene trap vectors contain a reporter gene without a functional promoter, while enhancer trap vectors carry a reporter gene preceded by a minimal promoter. The reporter gene is expressed in a fashion that reflects the normal expression pattern of the tagged gene (Groover *et al.*, 2004). A collection of poplar gene and enhancer trap lines has been established and is available for screening.[116]

TILLING is only just beginning to be applied in forest trees, and in hybrid poplar several induced mutations have been isolated by this method (Mattsson *et al.*, 2007). EcoTILLING, that detects natural mutations, has been used to catalog the level of DNA variation in natural populations of *P. trichocarpa*. With this technique, 63 novel SNPs were identified in nine target genes for 41 tree accessions (Gilchrist *et al.*, 2006). Such data can provide insights into gene function and be informative for association mapping analysis.

J. Transgenesis

FAO (2004) reports that forest tree genetic modification research takes place in 35 countries, with 48% of the reported activities occurring in North America, 32% in Europe, 14% in Asia, 5% in Oceania, and less than 1% each in Africa and South America (FAO, 2004). Field trials of GM trees have been restricted largely to four genera, that is, *Populus*, *Pinus*, *Liquidambar*, and *Eucalyptus*. To date, China is the only country where GM trees are reported to be commercially available, involving transgenic poplar trees grown on roughly 300–500 ha (FAO, 2004, 2011b).

V. CURRENT STATUS OF BIOTECHNOLOGIES FOR THE MANAGEMENT OF ANIMAL GENETIC RESOURCES

Nearly a billion people depend upon livestock for their livelihoods. Livestock provides income, food for human consumption, fiber, leather, fuel, draught power, and fertilizer, thus contributing to food security and nutrition. In recent years, there has been a surge in demand for livestock products especially meat and milk, termed the "livestock revolution," driven by continued population growth, rising affluence, and urbanization. Livestock is one of the fastest growing

[116]http://www.fs.fed.us/psw/programs/ifg/genetraps.shtml.

sectors of the agricultural economy and contributes 40% of the global value of agricultural output. Livestock also contributes 15% of total food energy and 25% of dietary protein at the global level (FAO, 2009c).

Both conventional technologies and biotechnologies have contributed to increased livestock productivity. However, the challenge is to maintain the diversity in animal genetic resources while simultaneously meeting the increasing demand for animal products. The use of appropriate biotechnologies, as described below, can play an important role in the management of animal GRFA for improved understanding of genetic diversity, enhanced conservation of breeds, increased animal productivity, and better disease management.

A. Molecular markers

Microsatellites are the most popular markers to estimate genetic diversity in livestock and have been widely used to assess within- and between-breed genetic diversity in cattle (Amigues *et al.*, 2011; Egito *et al.*, 2007; Freeman *et al.*, 2006; Kugonza *et al.*, 2011), pigs (Behl *et al.*, 2006; Sollero *et al.*, 2009), sheep (Peter *et al.*, 2007; Tapio *et al.*, 2010), goats (Cañón *et al.*, 2006; Xu *et al.*, 2010), chickens (Bodzsar *et al.*, 2009; Hillel *et al.*, 2003; Mtileni *et al.*, 2011), and horses (Leroy *et al.*, 2009; Ling *et al.*, 2010). However, information about genetic diversity in many indigenous livestock breeds is still scant.

With the intention of obtaining a global view of animal genetic diversity (by generating reproducible and comparable data and integrating national and regional datasets), panels of 30 microsatellite markers for nine major livestock species[117] have been recommended by FAO and the International Society of Animal Genetics (ISAG). To further standardize results, the development of SNP panels has also been proposed. In order to coordinate the various studies undertaken, guidelines on molecular genetic characterization have recently been developed by FAO[118] to aid countries in planning, implementing, and analyzing genetic diversity of their animal genetic resources.

Country reports prepared for the State of the World's Animal Genetic Resources indicate that in developing countries, genetic distance studies for livestock breeds have been undertaken in 4 countries in Africa, 6 in Asia, 11 in Latin America and the Caribbean, and 2 in the Near and Middle East (FAO, 2007a). Collaborative efforts between FAO, the Joint FAO/IAEA Division, and the International Livestock Research Institute are ongoing for molecular

[117] http://www.globaldiv.eu/docs/Microsatellite%20markers.pdf.
[118] http://www.fao.org/docrep/014/i2413e/i2413e00.pdf.

characterization of animal genetic resources in Asia and Africa (FAO, 2011c). The Joint FAO/IAEA Division has developed a web-linked database for sharing molecular data for the characterization of several species.[119]

Molecular characterization studies, using mitochondrial DNA, Y chromosomal, and autosomal variation, have provided insights into breed history and ancestral populations of many species, including cattle, sheep, goats, horses, pigs, and chickens (Groeneveld *et al.*, 2010). Molecular markers have also been informative for identifying the geographical site(s) of domestication such as the Near Eastern origin of modern European taurine cattle (Troy *et al.*, 2001) and the Indus valley origin of zebu cattle (Ajmone-Marsan *et al.*, 2010). Further, molecular marker data in cattle (Loftus *et al.*, 1999), goats (Cañón *et al.*, 2006), and sheep (Tapio *et al.*, 2010) have shown that breeds from near the putative domestication centers have higher levels of genetic diversity, suggesting that these breeds be prioritized for conservation, particularly in the first phase of conservation actions.

A major threat to conservation is introgressive hybridization, and therefore, early detection is fundamental for effective conservation strategies. Molecular markers have provided evidence for crossing of domestic pigs with European wild boars and of domestic cattle with wild North American bison (FAO, 2002; Halbert *et al.*, 2005). Interestingly, molecular marker data have also demonstrated gene flow from wild populations to domesticated animals, for instance, from junglefowl to domesticated populations of Vietnamese chicken (Berthouly *et al.*, 2009).

Development of breed-specific brands is a common approach for adding value to the products of local breeds, thus contributing to their continued survival in *in situ* conditions (Tixier-Boichard *et al.*, 2006). The price advantage obtained for the branded product can only be maintained if the uniqueness and purity of the breed origin can be demonstrated. Development of panels of molecular markers corresponding to breed-unique alleles has been proposed as a method to monitor the genetic source of such products and assure customers of their origin, thus protecting the market of the local breeds (Dalvit *et al.*, 2007).

Parentage determination is essential for accurate pedigree information, genetic evaluation, and successful breeding programs. Parentage analysis with DNA markers is much more precise and reliable than conventional testing with blood groups or biochemical marker systems, and incorrect parentage can be excluded with almost 99% accuracy (Glowatzki-Mullis *et al.*, 1995). The ISAG has suggested a panel of 12 loci each to be used in cattle (ISAG Conference, 2008a) and horse (ISAG Conference, 2008b) parentage analysis and the recommended loci have been effectively implemented for this purpose (Ozkan

[119]http://www.globalgenomic.com/.

et al., 2009; Seyedabadi *et al.*, 2006; Stevanovic *et al.*, 2010). Besides cattle and horses, parentage analysis employing microsatellites has been carried out in other livestock species (Araújo *et al.*, 2010; Siwek and Knol, 2010). SNPs have also been shown to be effective for parentage analysis (Hara *et al.*, 2010; Heaton *et al.*, 2002; Rohrer *et al.*, 2007).

Effective population size (N_e) is extensively used as a criterion for determining the risk status of livestock breeds, and molecular markers are commonly employed to estimate N_e, especially of natural populations. Proposed methodologies for estimating N_e from marker data have been reviewed in Wang (2005). Molecular marker-based methods have been used for the estimation of N_e in cattle (Flury *et al.*, 2010; Thévenon *et al.*, 2007), sheep (Álvarez *et al.*, 2008), horses (Goyache *et al.*, 2011), chickens (Márquez *et al.*, 2010), and pigs (Uimari and Tapio, 2011).

Molecular markers have been used to identify QTLs that affect traits of importance in livestock production, and a large number of studies have been carried out predominantly in developed countries (FAO, 2007b), since in the low-input systems that exist in many developing countries, the necessary phenotypic and pedigree information is often lacking. High-resolution linkage maps, to facilitate fine mapping of QTLs, have been created for livestock species such as chickens (Groenen *et al.*, 2009), cattle (Arias *et al.*, 2009), pigs (Vingborg *et al.*, 2009), and sheep (Raadsma *et al.*, 2009). Consolidated QTL data on multiple livestock species are publicly available on the Animal QTL database (AnimalQTLdb) for easily locating QTLs and making comparisons within and between species (Hu *et al.*, 2007). Currently, this database contains data on 6347 pig, 4802 cattle, 2451 chicken, and 639 sheep QTLs, representing 593, 382, 248, and 184 different traits, respectively.[120]

B. "Omics"

Whole genome sequencing of many domestic animals such as cattle, chickens, horses, pigs, sheep, and turkeys has been completed (Fan *et al.*, 2010, 2011; Stothard *et al.*, 2011) and has led to the identification of millions of SNPs in the major livestock species.[121] This increased genomic information has, in turn, facilitated candidate gene analysis (Rincón *et al.*, 2009; Seichter *et al.*, 2011). Comparative genomics studies with the sequences have yielded insights into the biology and evolution of animal species (International Chicken Genome Sequencing Consortium, 2004; The Bovine Genome Sequencing and Analysis Consortium, 2009; Wade *et al.*, 2009).

[120]http://www.animalgenome.org/cgi-bin/QTLdb/index.
[121]http://www.ncbi.nlm.nih.gov/SNP/.

Subsequent to whole genome sequencing, International HapMap (haplotype mapping) Projects have been developed. These are collaborative efforts for large-scale genotyping of a given species to identify haplotypes, that is, a set of SNPs on a single chromosome that are statistically associated (Cockett *et al.*, 2010; Groenen *et al.*, 2010; The Bovine HapMap Consortium, 2009). The resulting haplotype maps have facilitated the detection of selection signatures associated with domestication and breed formation (Qanbari *et al.*, 2010; Stella *et al.*, 2010).

Genome-wide association studies, based on high-throughput SNP genotyping technologies and combined with phenotypic data, are being used to explore loci associated with complex traits, for example, growth rate, milk production, and disease related traits and have the potential to lead to more efficient GS, by contributing to the understanding of biological mechanisms underlying complex traits (Fan *et al.*, 2010).

Transcriptomic approaches have provided mechanistic insights into numerous regulatory networks, for example, the host response to pathogen exposure (Rinaldi *et al.*, 2010; Tuggle *et al.*, 2010), biological pathways relevant to traits related to meat performance (Wimmers *et al.*, 2010), and ovarian follicle development as well as embryonic development for improving current-assisted reproductive technologies (Grado-Ahuir *et al.*, 2011; Huang *et al.*, 2010c). Genome-wide transcript profiling has also been undertaken for breed comparisons, such as the variation in parasite resistance within and among sheep breeds (MacKinnon *et al.*, 2009), differences in pork quality between pig breeds (Gao *et al.*, 2011), and hepatic expression of genes in selected lines of the Holstein-Friesian dairy cattle breed (McCarthy *et al.*, 2009).

Implementation of proteomic tools has assisted in understanding biological traits, particularly those that affect milk (Affolter *et al.*, 2010; D'Amato *et al.*, 2009; Wu *et al.*, 2010) and meat quality (Bjarnadottir *et al.*, 2010; Chaze *et al.*, 2008; Hornshoj *et al.*, 2009). Proteomic methods have also been applied in animal health to study the pathophysiology of diseases and to identify diagnostic markers for early detection of disease (Bendixen *et al.*, 2011). Proteomic strategies for characterizing breeds are just beginning to be exploited (Almeida *et al.*, 2010).

C. Bioinformatics

In addition to the previously mentioned databases, individual livestock databases also exist: for example, tools for genome annotation, discovery, and analysis are available for the bovine genome (Childers *et al.*, 2011). Other species-specific databases include the Chicken Variation database (ChickVD),[122] the Pig

[122]http://chicken.genomics.org.cn/index.jsp.

Genomic Informatics System (PigGIS),[123] and the International Sheep Genomics Consortium.[124] A comprehensive list of databases on livestock genomics is provided in Groeneveld *et al.* (2010).

Advances in bioinformatics are occurring at a very rapid pace, and consequently, there is a need for up-to-date training. The Biosciences eastern and central Africa (BecA)[125] Hub Bioinformatics Platform, a specialist node of the European Molecular Biology Network, provides advanced computational capabilities in bioinformatics and is involved in raising awareness and capacity development in the subject in Africa.[126] Training on bioinformatics tools is also provided by the Joint FAO/IAEA Division.[127]

D. Cryopreservation

Cryoconservation of animal genetic resources (gametes, embryos, somatic cells) has been carried out in a number of countries, including a few developing countries (FAO, 2007a). Gene banks have also been created for conserving rare livestock breeds (Long, 2008). However, the costs associated with collecting, cryopreserving, and reconstituting animal germplasm are comparatively much higher than those for plants. A recent survey undertaken by FAO in 90 countries illustrates that the number of cryoconservation programs is approximately half the number of *in situ* programs for most livestock species, with fully operational gene banks reported in only about 20% of the countries.[128]

Semen of most livestock species has been successfully cryopreserved with freezing procedures that are species-specific (FAO, 2007a). Cryoconservation of animal genetic resources in the form of semen is also practical due to its abundant availability and low cost. However, disadvantages include the conservation of only a single complement of chromosomes and the lack of mitochondrial genes. When only stored semen is used to reconstitute breeds by backcrossing with another breed (that is, when female gametes/live animals of the breed are completely lost), sufficient semen must be available for the required number of backcrosses, which can be several thousand doses for lowly reproductive species. Even with five or more generations of crosses, a nontrivial

[123]http://pig.genomics.org.cn/.

[124]http://www.sheephapmap.org/.

[125]BecA is an initiative developed within the framework of Centres of Excellence for Science and Technology in Africa and hosted and managed by the International Livestock Research Institute (ILRI).

[126]http://hub.africabiosciences.org/Bioinformatics.

[127]http://www-naweb.iaea.org/nafa/aph/prospectus-animal-genetics.pdf.

[128]http://www.fao.org/docrep/meeting/021/am132e.pdf.

proportion of the genetic material of the backcrossed breed will remain (Boettcher *et al.*, 2005). Moreover, in avian species, the W chromosome is absent in males and thus semen cryopreservation cannot be used to conserve the genes on that chromosome (FAO, 2007a).

Due to low permeability to cryoprotectants, cryopreservation of oocytes remains a challenge and is not as well established as for semen and embryos. Nevertheless, significant progress has been made and viable oocytes have been recovered after freezing and thawing in many mammalian species (Critser *et al.*, 1997; Dhali *et al.*, 2000). Offspring born from embryos produced from cryopreserved oocytes have been reported in cattle, sheep, and horses (Prentice and Anzar, 2011). Oocyte cryopreservation of avian species has not been successful due to the large size and high lipid content. Oocyte storage requires complementary cryoconservation of semen if the genetic material on the Y chromosome of mammalian species is not to be lost.

Embryo cryoconservation allows the conservation of the full genetic complement but is more expensive and requires greater technical capacity than semen cryoconservation (Gandini *et al.*, 2007). The success of cryopreservation depends on the origin and developmental stage of the embryos, with *in vivo* derived embryos withstanding cryopreservation better than *in vitro* produced embryos and especially good results being attained with blastocysts. Cryopreservation of embryos has been reported for virtually all of the major mammalian livestock species, though its widespread use is limited to cattle, sheep, and goats (Prentice and Anzar, 2011). Cryopreservation of embryos from pigs and equine species has been quite problematic due to their extreme chilling sensitivity, but in recent years, pig embryos have been successfully cryopreserved (Vajta, 2000) and live piglets have been obtained from cryopreserved pig embryos (Nagashima *et al.*, 2007). Live offspring, from cryopreserved embryos, have also been obtained in horses (Ulrich and Nowshari, 2002) and other livestock species (Paynter *et al.*, 1997; Rodriguez-Dorta *et al.*, 2007).

Somatic cells can be collected rapidly at favorable costs and easily cryopreserved (Groeneveld *et al.*, 2008). However, utilizing the stored material through reproductive somatic cell nuclear transfer is more complex and expensive, and mitochondrial genes are lost.

The choice of genetic material for cryoconservation depends on the generation interval and reproductive rate of the species, and costs must also be considered. For example, embryo collection and freezing in livestock is much more expensive than for semen, but regeneration using embryos is quicker and cheaper. These differences can vary greatly by species. The use of semen to regenerate a breed of cattle or horse would take much longer and require many more doses of semen than for pigs or rabbits, whereas with embryos, the regeneration time could be less than a year and the numbers of embryos required would be similar for all species. Draft guidelines providing technical advice on the cryoconservation of animal genetic resources have been developed by FAO.[129]

E. Reproductive biotechnologies

In livestock, both AI and ET can be applied for the future use of cryopreserved GRFA. Once thawed, the semen can be used for backcrossing using AI, while the embryos can be transferred to recipient females. AI and ET have also been used to evaluate the viability of cryopreserved germplasm, subsequent to long-term storage. Pure-bred beef cows inseminated with frozen-thawed Angus bull semen, processed during three time periods (from the 1960s to 2002), resulted in similar pregnancy rates across the different time periods, demonstrating that good-quality semen frozen in liquid nitrogen should remain viable indefinitely (Carwell *et al.*, 2009). Fogarty *et al.* (2000) showed that cryopreserved sheep embryos, stored for 13 years, could be successfully thawed and transferred to recipient ewes.

AI can have both positive and negative effects on the sustainable use and diversity of animal genetic resources. It has allowed for the tremendous increases in productivity, perhaps increasing the economic sustainability of livestock production, but has contributed to decreased effective population sizes and thus diversity. AI, together with cryoconservation, has facilitated the transboundary gene flow of animal genetic resources, largely in the North-South direction, which has threatened local populations (along with the lack of genetic improvement programs for local breeds). When AI is applied in the absence of accurate genetic evaluation, it can greatly decrease genetic diversity without yielding gains in productivity. ET can have similar positive or negative effects, but these tend to be less because of the technology's relatively higher cost and lower numbers of offspring per animal.

1. Artificial insemination

AI is the most widely used reproductive technology in developed and developing countries. It has the best cost–benefit among all reproduction technologies and has revolutionized the animal breeding industry. It enables a single bull to be used simultaneously in several countries for many inseminations a year and also enhances the efficiency of progeny testing of bulls. AI is applied in cattle, sheep, goats, turkeys, pigs, chickens, and rabbits, with its most extensive application being in cattle. It is especially indispensable in the turkey breeding

[129]http://www.fao.org/docrep/meeting/021/am136e.pdf.

industry since the large size of the males of broad-breasted turkeys, which have been bred for body conformation, precludes natural mating (Flint and Woolliams, 2008).

As per 2002 statistics, more than 100 million AIs in cattle, 40 million in pigs, 3.3 million in sheep, and 0.5 million in goats are performed globally every year (Thibier and Wagner, 2002). In developing countries, most of the AI services are provided by the public sector, though it is still unavailable in many countries in Africa and the Southwest Pacific (FAO, 2007a). When applied, in Africa, Asia, Latin America, and the Caribbean, AI is mostly used in cattle (especially in the dairy sector), with semen from exotic breeds predominating for local livestock production (FAO, 2011c).

AI is generally not very expensive and can be carried out by trained livestock keepers. Even so, the effectiveness of AI is often limited by organizational and logistical constraints in developing countries. Moreover, for AI to contribute to increased rural livelihoods, it has to be complemented by other services to maintain the health and fertility of the inseminated animals and provide enhanced market access for the product.

2. Progesterone monitoring

Progesterone monitoring can be used to verify whether animals have been inseminated at the correct time and detect any animals that later return to estrus so that they can be reinseminated without delay (Dargie, 1990). The Joint FAO/IAEA Division has promoted the development and transfer of progesterone RIA (based on ^{125}I) in about 30 developing countries in Asia, Africa, and Latin America for improved livestock productivity (FAO, 2011c).

3. Estrus synchronization

The administration of exogenous reproductive hormones to synchronize ovulation is most often used in cattle, sheep, and goats, although the use of estradiol and its related ester derivatives is prohibited in many countries due to concerns about the effects of steroid hormones in the food chain (Lane *et al.*, 2008; Lucy *et al.*, 2004). In addition to increasing AI efficiency, estrus synchronization is also of utility for optimizing transfer pregnancy rates when ET is carried out between donor and recipient females. Estrus synchronization in developing countries is restricted to intensively managed farms or smaller farms with links to farmers' associations that routinely use AI (FAO, 2011c).

4. Embryo transfer

ET enhances the ability to select superior female genetic material, enables low cost transportation of diploid genetic material, and allows for more offspring to be produced from a single female animal. With the intention of ensuring that specific pathogens that could be associated with embryos are controlled and transmission of infection to recipient animals and progeny is avoided, the World Organiztion for Animal Health (OIE) has recommended measures on the collection and processing of *in vivo* derived embryos[130] and *in vitro* produced embryo/oocytes[131] from livestock and horses. The OIE has also provided recommendations for the collection and processing of micromanipulated embryos/oocytes (that is, subjected to biopsy, nuclear transfer, etc.) from livestock and horses.[132]

ET has been applied in many species, including cattle, buffaloes, horses, pigs, sheep, and goats. The technology is most widely used in cattle, with 535,164 *in vivo* derived bovine embryos reported to have been transferred in 2009 compared to 352 transfers in goats (down from 20,000 in 2006) and 24,470 in horses worldwide (Stroud, 2010). North America has been the center of commercial ET activity, with 46% of all reported *in vivo* derived bovine ETs in the world, followed by Europe with 18% of the total. More frozen *in vivo* derived embryos were transferred in 2009 than fresh embryos except in South America, where four times as many fresh embryos than frozen embryos were transferred. The number of *in vitro* produced bovine embryos transferred increased by 17% from 2008 to 2009, with South America responsible for 84% of the transfers (Stroud, 2010).

In developing countries, ET is reported to be applied in 5 countries in Africa, 8 in Asia, and 12 in Latin America and the Caribbean (FAO, 2007a). ET is expensive and requires highly skilled personnel which explains, in part, its low level of use. Compared with natural breeding, the cost of breeding through multiple ovulation and embryo transfer has been estimated to be over 60 times higher (Maxwell and Evans, 2009). A recent study on ET in Mexico showed that the technology is profitable for farmers only when substantial subsidies are provided (Alarcon and Galina, 2009). Once the subsidized initiative ends, the

[130]http://www.oie.int/fileadmin/Home/eng/Health_standards/tahc/2010/en_chapitre_1.4.7.pdf.

[131]http://www.oie.int/fileadmin/Home/eng/Health_standards/tahc/2010/en_chapitre_1.4.8.pdf.

[132]http://www.oie.int/fileadmin/Home/eng/Health_standards/tahc/2010/en_chapitre_1.4.9.pdf.

high costs associated with preparing the donors and recipients, embryo recovery and transfer, and the gestation itself make the ET program unfeasible for farmers to sustain on their own.

5. *In vitro* fertilization

IVF has the potential to accelerate genetic progress by enhancing the accuracy and intensity of selection, reducing the generation interval, and improving pregnancy rates in herds with low fertility (Hansen, 2006). IVF may reduce inbreeding since it offers greater flexibility in the mating design of sires to cows. IVF also allows for more efficient utilization of sperm since, compared to AI, a relatively lower number of spermatozoa can produce viable embryos (Boa-Amponsem and Minozzi, 2006). This is particularly valuable when small amounts of sexed sperm are available.

IVF with frozen sperm is mainly applied in commercial cattle ET in developed countries and is not yet routinely used in developing countries due to its high cost and requirements for a well-equipped laboratory and skilled technicians. However, this technology has particularly expanded in Brazil and Japan, which account for the majority of *in vitro* fertilized embryos. In 2009, the total number of transferable bovine *in vitro* produced embryos was 379,000, with Brazil responsible for 68%, followed by Japan with 20% of the world's *in vitro* embryo production (Stroud, 2010). Live offspring from *in vitro* fertilized embryos have also been reported in sheep, goats, and pigs.[133]

Recently, attempts have been made to utilize IVF procedures for overcoming reproductive barriers between species and producing hybrids. Owiny *et al.* (2009) demonstrated that although domestic cattle oocytes could be fertilized by African buffalo sperm, the resulting embryo development was slow and low, most likely due to chromosomal disparity.

6. Cloning

Another reproductive technology that can be used for conservation purposes, particularly when a breed is nearly extinct, is that of cloning. Cloning also offers the opportunity of increasing the uniformity of a given product and dissemination of superior genotypes in commercial populations. Somatic cell nuclear transfer has been successful in cattle, sheep, goats, pigs, rabbits, camels, and horses, but the efficiency is low and often cloned offspring show abnormalities (Kues and Niemann, 2004).

Although cloned animals have been produced in a few developing countries (e.g., in China and India), it is still at the experimental stage due to the high costs and skills required (FAO, 2011c). According to a survey carried

[133]http://www.fao.org/docrep/meeting/021/am136e.pdf.

out by OIE in 2005 (with 91 respondent countries), cloning capabilities were reported by 4% of the countries in Africa, 23% in the Americas (Latin America and the Caribbean and North America), 23% in Asia, and 18% in Europe (MacKenzie, 2005).

Nevertheless, developments in animal cloning have made realistic the conservation of animal genetic resources through cryopreservation of somatic cells rather than germ cells. This strategy can significantly decrease the costs and the level of technical expertise required to collect and bank the genetic material, but relies, for most species, on the assumption that the use of the material for regeneration of new animals will not be necessary until future technological advances have increased the efficiency and reduced the cost and animal welfare + +implications of creating clones. From a regulatory perspective, the use of cloned animals for food production has been approved by only a few national governments.

7. Sexing

Sperm and embryo sexing techniques allow for the sex of the progeny to be predetermined, which is particularly useful for sex-limited and sex-influenced traits (e.g., for cattle, females are desired as dairy animals and males as beef animals).

Sexing of sperm can effectively increase selection intensity within dam[134] pathways, enhancing genetic response for traits associated with productivity or other contributors to sustainability. Successful application of sperm sexing has been limited due to the high cost of sexed semen as well as low sperm viability and fertility rates. Consequently, much of the research on sperm sexing has been conducted by the private sector. Sperm sexing in bovine species has been applied commercially in a few countries, including Argentina, Brazil, Canada, Denmark, the Netherlands, the United Kingdom, and the United States of America, and approximately 2 million calves have been produced worldwide from insemination with sexed sperm since 2000 (Rath, 2008). In other livestock species, the technology has not yet reached a stage that allows for commercial application (Rath and Johnson, 2008).

Embryo sexing requires the removal of a small number of cells from the embryo in order for the assay to be performed. Embryo biopsy is thus an invasive technique that calls for a high level of technician skill. The additional costs for embryo biopsy and sexing as well as the increased sensitivity of biopsied embryos to freezing/thawing procedures compared with intact ones are some of the factors that have limited the application of this technique particularly in developing countries (Heyman, 2010).

[134]Female parent of an animal, especially domestic livestock.

F. Biotechnologies for disease diagnosis and prevention

Infectious animal diseases cause devastating losses in the livestock industry, limit efficient production, and, in the case of zoonotic diseases, pose threats to human health. Livestock farmers have suffered severe economic losses due to major outbreaks of transboundary animal diseases (TADs) such as foot and mouth disease (FMD) in Europe, classical swine fever (CSV) in the Caribbean and Europe (1996–2002), rinderpest (RP) in Africa in the 1980s, peste des petits ruminants (PPR) in Bangladesh and India, contagious bovine pleuropneumonia (CBPP) in Eastern and Southern Africa (late 1990s), as well as Rift Valley fever in the Arabian Peninsula (2000) (Domenech *et al.*, 2006).

In recent years, FAO has collaborated with various international and regional organizations for preventing, controlling, and managing TADs. The Global Framework for Progressive Control of Transboundary Animal Diseases (GF-TADs), a joint FAO/OIE initiative, provides a framework to address endemic and emerging infectious diseases. The FAO-OIE-WHO Global Early Warning and Response System for Major Animal Diseases, including Zoonoses,[135] avoids duplication and coordinates the verification processes of the three organizations to improve the early warning and response capacity to animal disease threats. The objective of the multiagency One World One Health Strategic Framework[136] (FAO–OIE–WHO–UNICEF in collaboration with the World Bank and the UN System Influenza Coordinator) is to diminish the risk and minimize the global impact of epidemics and pandemics due to emerging infectious diseases at the animal–human–ecosystems interface.

Diagnostics and vaccines based on recombinant DNA technology have immensely contributed and are increasingly being used for improving disease control strategies (Balamurugan *et al.*, 2010; Barnard, 2010). The OIE has been instrumental in providing updated internationally agreed diagnostic laboratory methods and requirements for the production and control of vaccines for the diseases listed in the OIE *Terrestrial Animal Health Code* (OIE, 2010b).

1. Diagnostics

Accurate, rapid, and early disease diagnosis allows effective control measures to be implemented. Improved, robust real-time RT-PCR assays have been developed for the detection of viral pathogens that cause FMD, CSV, bluetongue disease, avian influenza, and Newcastle disease (Hoffmann *et al.*, 2009).

[135] http://www.glews.net/.

[136] http://www.aitoolkit.org/site/DefaultSite/filesystem/documents/OWOH_14Oct08.pdf.

The various ELISA- and PCR-based methods that are being used and are recommended to diagnose diseases of global importance are described in the OIE *Manual of Diagnostic Tests and Vaccines for Terrestrial Animals* (OIE, 2010b). The OIE has also adopted a Twinning Program,[137] that is, a partnership between OIE Reference Laboratories[138] and candidate laboratories to develop laboratory diagnostic methods based on the OIE Standards in developing and transition countries.

Assays may be applied for many purposes (e.g., to document freedom from infection in a defined population, prevent spread of disease through trade, eradicate a disease from a region or country, confirmatory diagnosis of suspect or clinical cases, etc.) and thus need to be validated for their intended purpose. The OIE has developed criteria that must be fulfilled during assay development and validation of all assay types.[139]

The Joint FAO/IAEA division has been at the forefront of developing and delivering early and rapid diagnostic kits to developing countries, as well as providing laboratory networking and on-site training to detect infectious disease agents.[140] Seventy countries use disease diagnostic tests developed or validated by the FAO/IAEA Joint Division to assist their animal disease management programs (FAO/IAEA, 2008). Currently, efforts are underway to develop pen-side tests utilizing the LAMP PCR for the diagnosis of avian influenza, CBPP and PPR, that can be applied in rural and difficult to reach remote areas (IAEA, 2010).

The Global Rinderpest Eradication Program,[141] spearheaded by FAO and in close association with OIE, was conceived to promote the global eradication of, and verification of freedom from, the disease. No known RP outbreaks have been detected since 2001 and its eradication from all 198 countries of the world was announced in 2011, a milestone in veterinary history.[142] In addition to the development of an improved vaccine against RP, biotechnologies such as C-ELISA and RT-PCR played a vital role in disease diagnosis and surveillance, including for the differential diagnosis between the closely related RP and PPR viruses. Rapid diagnosis, to prevent further spread of disease, was made possible by developing a field-based penside test using eye swab materials.

[137]http://web.oie.int/downld/LABREF/A_Guide.pdf.

[138]http://web.oie.int/eng/normes/mmanual/2008/pdf/XX_LIST_LAB.pdf.

[139]http://web.oie.int/eng/normes/mmanual/2008/pdf/1.1.04_VALID.pdf.

[140]http://www-naweb.iaea.org/nafa/aph/public/aph-nl-52.pdf.

[141]http://www.fao.org/ag/againfo/programmes/en/grep/home.html.

[142]http://www.fao.org/news/story/en/item/79335/icode/.

2. Vaccines

Vaccination is a cost–effective way to prevent clinical signs of a disease after infection or to help manage and even eradicate an infection. Ensuring vaccine purity, safety, potency, and efficacy is indispensable for producing high-quality vaccines and the OIE has described the principles of veterinary vaccine production (OIE, 2010b). However, vaccines must complement other aspects of disease management, for instance, surveillance and diagnostic proficiency.

Gene-deleted vaccines have been licensed for use against bacterial diseases in horses, chickens, turkeys, and sheep (Meeusen *et al.*, 2007). Several commercial veterinary vaccines based on viral vectors (e.g., DNA viruses such as poxviruses, herpesviruses, and adenoviruses) have been developed, while many RNA viruses (e.g., Newcastle disease virus and a few retroviruses) are currently being evaluated as vectors (Gerdts *et al.*, 2006). Rapid progress is also being made in developing bacteria as vectors (Rogan and Babuik, 2005).

DNA vaccines have been shown to be effective in cattle, sheep, and poultry but have been commercially produced only for West Nile virus in horses (Potter *et al.*, 2008). Recombinant subunit vaccines against porcine circovirus type 2 (Fachinger *et al.*, 2008) as well as atrophic rhinitis (OIE, 2010a) have been commercialized, while in recent years, efforts to develop orally delivered plant-based subunit vaccines have emerged (Streatfield and Howard, 2003; Yusibov *et al.*, 2011).

Vaccines against economically devastating livestock diseases such as FMD and CSV are not used in disease-free countries as they would interfere with disease surveillance and compromise export interests. In order to address the ethical, welfare, and economic aspects of large-scale animal culling in response to outbreaks of these infectious diseases, DIVA vaccines together with their companion diagnostic tests are now available (Beer *et al.*, 2007; Grubman, 2005). Furthermore, vaccine banks have been established to guarantee supplies for emergency vaccination in case of outbreaks.

Vaccine banks may either store a ready-to-use formulated vaccine with a short shelf life, or more commonly the antigen component, that can be stored for a long period, for subsequent formulation into vaccine when required. The OIE has published guidelines for international standards for vaccine banks.[143]

G. Molecular marker-assisted selection

Molecular markers can be used to enhance within-breed selection through MAS. In addition to reduced generation interval and more accurate selection, MAS in livestock can be especially advantageous for traits expressed in one sex, traits that are expressed very late in life as well as traits that usually necessitate slaughtering some of the animals.

[143]http://web.oie.int/eng/normes/mmanual/2008/pdf/1.1.10_VACCINE_BANKS.pdf.

MAS has been implemented in large-scale dairy cattle breeding schemes in developed countries (Bennewitz *et al.*, 2004; Boichard *et al.*, 2002), but its application has been limited in developing countries (FAO, 2007b). Examples of gene or marker tests used in commercial breeding for different species are reviewed in Dekkers (2004). For MAS to succeed and be sustainable within smallholder production systems in developing countries, other system constraints such as animal management, government policies, marketing, etc., need to be addressed (Marshall *et al.*, 2011).

The huge variety across breeds can be exploited for genetic improvement through marker-assisted introgression, that is, by introgressing individual genes or QTLs from one breed into another through repeated backcrossing. The Booroola gene (*FecB* gene), for enhancing prolificacy, has been introgressed by cross-breeding from the Booroola Merino into Awassi and Assaf breeds in Israel (Gootwine *et al.*, 2001, 2003) and from the small Garole breed into the more productive but lowly fecund Deccani breed in India (Nimbkar *et al.*, 2007).

GS is fast becoming the new paradigm in animal breeding. Since the genomic breeding value can be predicted at birth, the accuracy of prediction for young animals can be increased. Improved accuracy of predicting breeding values leads to increased genetic gain without increasing the rate of inbreeding (Daetwyler *et al.*, 2007). It has also been demonstrated that cross-bred populations can be used as a training data set for GS for predicting breeding values of pure-bred animals without a substantial loss of accuracy compared with training on pure-bred data, thus eliminating the need to track pedigrees (Toosi *et al.*, 2010).

GS has the potential to increase response to selection, but phenotype and pedigree recording systems need to be in place. This approach is being used in developed countries, particularly for dairy cattle (Hayes *et al.*, 2009) and more recently for chickens (Chen *et al.*, 2011), but has high barriers for entry (financial, technical, and informational) that have till now precluded its application in developing countries. GS is thus likely to increase the gap in genetic merit for product yield between local and international transboundary breeds but may also limit the adaptability of these latter breeds to harsh environments, if adaptability is adversely correlated with production.

H. Mutagenesis

In the livestock sector, mutagenesis is generally not applied to animal populations. However, the sterile insect technique (SIT), an environment-friendly method for the management of insect pests of agricultural and veterinary importance, has proved to be an effective technique when used as part of an area-wide integrated pest management (AW-IPM) approach (Vreysen and Robinson, 2010). The SIT relies on the introduction of sterility in wild female insects of the pest population when they mate with released radiation-sterilized males and has been used by 30 countries for the suppression/eradication of key insect pests (FAO/IAEA, 2008). Aspects of the SIT that make it a distinctive integrated pest management (IPM) tool include species specificity toward the target pest population as well as increased efficiency with decreasing target population density. As the SIT is extremely species specific, it has limited efficacy for the control of a broader range of vector species.

The most successful SIT campaign eradicated New World screwworm (which causes myiasis in warm-blooded animals, leading to loss of milk, meat, or wool production) from Central America, some islands in the Caribbean, Mexico, Panama, and southern United States of America, and the area is maintained screwworm-free through the weekly release of 40 million sterile flies in Panama to prevent reinvasion from South America (Robinson *et al.*, 2009). The first report of the New World screwworm occurring outside the Americas was in the Libyan Arab Jamahiriya in 1988. FAO, together with IAEA, the International Fund for Agricultural Development and the United Nations Development Program, was successful in containing the disease by dispersing sterile pupae and the country was declared officially screwworm-free in 1992. The annual producer benefits of the eradication program have been estimated to be USD 796 million, USD 292 million, and USD 77.9 million in the United States of America, Mexico, and Central America, respectively, with an estimated benefit/cost ratio of 5:1 in the infested zone in the Libyan Arab Jamahiriya (Vargas-Terán *et al.*, 2005).

Old World screwworm fly has a very wide distribution throughout Asia, tropical and subtropical Africa, the Indian subcontinent, parts of the Middle East, and Papua New Guinea. Small-scale field trials have shown the technical feasibility of using the SIT in the control of this pest and efforts are underway to use the technique as part of an AW-IPM approach on a much wider scale (Robinson *et al.*, 2009).

Trypanosomosis is a severe disease transmitted by tsetse flies in sub-Saharan Africa. When left untreated, it is fatal in livestock and if not lethal, reduces fertility, weight gain, and meat and milk offtake by at least 50% (Feldmann *et al.*, 2005). The effectiveness of the SIT in creating tsetse-free zones has been demonstrated in Zanzibar (1994–1997), where the tsetse population of *Glossina austeni* was completely eradicated from Unguja Island. Socio-economic assessments after the eradication indicated an increase in the contribution of the livestock sector to agricultural GDP (from 12% in 1986 to 34% in 1997) and in domestic food production leading to a decrease in food imports. An increase in average income per month of farming households, of 30% from 1999 to 2002, was also observed (Feldmann *et al.*, 2005). Following this success and the escalating incidence of trypanosomosis, renewed efforts to control tsetse in Africa culminated in the establishment of the Pan African

Tsetse Eradication Campaign,[144] a concerted initiative that uses conventional methods in combination with the SIT to eliminate tsetse and trypanosomosis from Africa.

I. Transgenesis

No transgenic livestock have been commercialized as food to date, though a number of transgenic animals with medical or bio-"pharming" applications are at different stages of commercial development (FAO, 2011c). Transgenic production capability has been reported by 8% of the countries in Africa, 15% in the Americas (Latin America and the Caribbean and North America), 23% in Asia, and 26% in Europe (MacKenzie, 2005).

VI. CURRENT STATUS OF BIOTECHNOLOGIES FOR THE MANAGEMENT OF AQUATIC GENETIC RESOURCES

The fisheries and aquaculture sector provides income and livelihood for almost 45 million people globally, and employment in this sector has grown faster than the world's population and employment in traditional agriculture (FAO, 2010a). In 2008, 115 million tons of fish was consumed as human food; fish supplied over 3 billion people with at least 15% of their average animal protein intake (FAO, 2010a). Aquaculture accounts for 46% of the world's food fish and is expected to overtake capture fisheries in this regard. Currently, the most caught species at the global level is the anchoveta, while carps are the most cultured group of species in the world (FAO, 2010d).

In contrast to the crop and livestock sectors, aquatic genetic resources comprise a very large number of species. Most fisheries harvest wild populations and, with a few exceptions, the majority of aquaculture species have short histories of domestication and thus are genetically much closer to their wild counterparts. Structured selective breeding programs have also been recently established and have been extremely effective for improving a range of characters for enhanced production (Bartley *et al.*, 2009). Consequently, the fisheries and aquaculture sector has made less use of biotechnologies compared to the crop and livestock sectors. Nevertheless, as reviewed in this section, the application of biotechnologies for the characterization, conservation, and

[144] http://www.africa-union.org/root/au/AUC/SpecialProjects/pattec/pattec.htm.

utilization of aquatic genetic resources is steadily increasing, especially in light of the increasing demand for fish and the need for the sustainable management of these resources.

A. Molecular markers

Intraspecific characterization studies using isozymes and microsatellites have demonstrated that individual marine species generally exhibit lower levels of interpopulation differentiation than freshwater species, since there are fewer barriers to migration and gene flow (Grant, 2006; Primmer, 2006). Molecular markers have been used to document the intraspecific population genetic structure and patterns of diversity in many natural populations, such as neotropical marine and freshwater fish in Latin America (Oliveira *et al.*, 2009), coho salmon in North America (Beacham *et al.*, 2011), Pacific cod in the Republic of Korea (Kim *et al.*, 2010a), and native tilapia in East Africa (Angienda *et al.*, 2011), among others. Genetic monitoring of natural populations over time can be applied to design effective conservation and management strategies (Sønstebø *et al.*, 2007; Van Doornik *et al.*, 2011).

In commercial capture fisheries, the catch often consists of a mixture of distinct populations and stock analysis is important to assess the relative contribution of each population in order to avoid overexploitation of some populations and optimize harvesting of abundant populations. Mixed stock analysis, based on molecular data, has been utilized to monitor stocks of many species including Atlantic salmon (Griffiths *et al.*, 2010), Pacific salmon (Flannery *et al.*, 2010), Atlantic cod (Wennevik *et al.*, 2008), lake sturgeon (Bott *et al.*, 2009), and steelhead trout (Winans *et al.*, 2004). The ability to link an individual fish product (e.g., shark fins and caviar) to the population of origin using molecular data is a robust tool for trade surveillance since it can be used to identify stocks that are major contributors to the market, although it requires prior sampling of the potential source populations (Baker, 2008; Chapman *et al.*, 2009).

Molecular markers have also been employed to distinguish the various species that occur together in mixed catches and are difficult to differentiate morphologically (Berntson and Moran, 2009; Okumus and Ciftci, 2003). Molecular markers are valuable as forensic tools, when consumer fraud or illegal possession of species at risk is suspected, especially for identifying early life stages, that is, eggs and larvae, and morphologically unrecognizable fish products (Ogden, 2008). Applications range from the detection of mislabeled products from threatened or poisonous species (Teletchea, 2009) to the investigation of illegally traded caviar from endangered sturgeon species (Ludwig, 2008) or the discrimination of Atlantic wolffish from the threatened spotted and northern wolffishes, which require legal permits for possession (McCusker *et al.*, 2008). Additionally, DNA arrays (based on the mitochondrial gene 16S rRNA) that have the capability to distinguish fish species, including eggs, larvae, and processed products, have been developed for consumer protection and trade regulation (Kochzius *et al.*, 2008, 2010).

Misidentification of cryptic species, that is, discrete species that are difficult or sometimes impossible to distinguish morphologically but are genetically distinct, can have serious negative consequences in fisheries management (Bickford *et al.*, 2007). Molecular markers have helped in cryptic species recognition across a wide variety of marine species, including finfish (Kon *et al.*, 2007; Piggott *et al.*, 2011), molluscs (Vrijenhoek, 2009), and crustaceans (Belyaeva and Taylor, 2009), suggesting that they are vital for accurately describing species diversity. DNA barcoding, based on a fragment of the mitochondrial gene *cytochrome c oxidase subunit 1*, is being applied for the identification of cryptic species as well as new and invasive species, and for regulatory and enforcement activities (Kim *et al.*, 2010b; Radulovici *et al.*, 2010; Wong *et al.*, 2011). The Fish Barcode of Life Initiative (FISH-BOL)[145] aims to barcode all the world's fishes (Ward *et al.*, 2009) and has currently barcoded 8293 species.

In aquaculture, molecular markers have provided information on genetic variability within and between hatchery stocks (Blackie *et al.*, 2011; Freitas *et al.*, 2007). It is particularly imperative to maintain high levels of genetic diversity in broodstocks used for replenishing and maintaining the viability of depleted fishery populations and also to ensure the productivity of genetically improved stocks for fish production. Concerns have been raised about the possible effects of stock enhancement on natural populations as molecular studies comparing hatchery and wild stocks have shown that farmed populations often have reduced genetic diversity and a lower effective population size (Araki and Schmid, 2010).

For an endangered species, *Brycon insignis*, a 28–57% reduction in allelic richness was detected in the broodstock used for restocking purposes compared to the wild populations, highlighting the need to maintain genetic variability in captive populations (Matsumoto and Hilsdorf, 2009). Similarly, in contrast to wild populations, lower genetic diversity has been observed in some hatchery populations, for example, Atlantic salmon (Blanchet *et al.*, 2008), Japanese flounder (Shikano *et al.*, 2008), and white seabream (Pereira *et al.*, 2010). At the same time, comparable levels of genetic diversity between wild and hatchery populations have also been reported (Pan and Yang, 2010).

Molecular markers have been used to investigate the impact of stock enhancement on wild populations (Okumus and Ciftci, 2003; Povh *et al.*, 2008) and to manage broodstocks for stock enhancement and restocking purposes (Congiu *et al.*, 2011). Microsatellite data showed that the genetic diversity of Chinook salmon was maintained over multiple generations of supplementation,

[145]http://www.fishbol.org/.

over a period of 16 years, with increased effective population sizes of both the wild and captive-origin adults (Eldridge and Killebrew, 2008). Conversely, when a captive broodstock was used for extensive release of red sea bream for 30 years, the disappearance of some rare alleles in the wild populations was detected (Kitada *et al.*, 2009).

The most common application of molecular markers (particularly microsatellites) in aquaculture species has been in parentage analysis in order to manage the rate of inbreeding in conventional breeding programs. Molecular markers are especially useful when physical tagging is not feasible or when individuals are reared communally to minimize environmental variation, as they can retrospectively assign individuals to families after evaluation of individual performance (FAO, 2011d). Parentage analysis with molecular markers has been successfully accomplished in many commercially important species, including Atlantic salmon (Norris *et al.*, 2000), Atlantic cod (Herlin *et al.*, 2008), gilthead sea bream (Brown *et al.*, 2005), Senegal sole (Castro *et al.*, 2006), common carp (Vandeputte *et al.*, 2004), Japanese flounder (Hara and Sekino, 2003), shrimp (Jerry *et al.*, 2006), and mussels (MacAvoy *et al.*, 2008), although its potential has not yet been fully exploited in developing countries.

Molecular analysis of parentage has also been utilized in natural populations for increased understanding of mating systems, for example, for detecting multiple mating in both sexes in Atlantic salmon (Wilson and Ferguson, 2002), multiple paternity in shrimp, crayfish, crab, and lobster (Yue and Chang, 2010; Yue *et al.*, 2010), and brood parasitism in cichlids and sunfish (DeWoody and Avise, 2001; Taborsky, 2001). Parentage analysis has been used to evaluate patterns of larval dispersal to determine connectivity among marine populations (Christie *et al.*, 2010; Planes *et al.*, 2009). Recently, this approach was employed to infer individual movement throughout the life cycle of the stream-dwelling brook charr, especially at very early life stages when individuals are too small and numerous to be tagged by other means (Morrissey and Ferguson, 2011).

Molecular markers are very effective for identifying farmed escapees, the origin of these escapees, as well as for quantifying the interactions between escaped farmed and wild fish. Physically tagging farmed fish is expensive, requires highly developed logistic systems for marking and tracking the fish, and has animal welfare implications. Molecular tagging, on the other hand, can circumvent these issues and has been successfully applied for identifying escaped fish (Glover, 2010). Molecular studies have documented introgression of farmed escapees with wild fish, leading to loss of local adaptation (Bekkevold *et al.*, 2006; Skaala *et al.*, 2006).

To facilitate molecular characterization, the Network of Aquaculture Centres in Asia-Pacific (NACA) has produced a two-part manual on the application of molecular tools in aquaculture and inland fisheries management, with part 1 focusing on the conceptual basis of population genetic approaches (Nguyen *et al.*, 2006a) and part 2 providing laboratory protocols and methodologies for data analysis and project design (Nguyen *et al.*, 2006b).

In aquaculture, QTL mapping using commercial populations has been accomplished mainly in developed countries (FAO, 2011d) and, in general, is not as advanced as in the crops and livestock sectors. Genetic linkage maps for QTL identification have been created for various fish and shellfish species such as Atlantic salmon, rainbow trout, channel catfish, tilapia, Japanese flounder, common carp, turbot, sea bass, sea bream, shrimp, oysters, and mussels (Castaño-Sánchez *et al.*, 2010; FAO, 2007b, 2011d; Liu, 2007). A number of QTLs for important traits have been mapped in farmed aquatic species, for instance, growth, flesh color, spawning time, sex determination, abiotic stress tolerance, and disease resistance (Baranski *et al.*, 2010; FAO, 2007b; Liu and Cordes, 2004; Loukovitis *et al.*, 2011; Presti *et al.*, 2009). Furthermore, aquaculture, while still relatively in its initial stages, can benefit from the large amount of information available for livestock species, especially for the identification of genes homologous to putative candidate genes such as those correlated with growth (De-Santis and Jerry, 2007).

Molecular markers have additional applications in aquaculture. For example, they are routinely applied to assess desired genetic manipulations, such as triploidy and gynogenesis (Jenneckens *et al.*, 1999; Presti *et al.*, 2009).

B. "Omics"

The genomes of five well-known model finfish species, the zebrafish, medaka, spotted green pufferfish, Japanese pufferfish, and three-spined stickleback (Oleksiak, 2010) and some marine invertebrates, for example, purple sea urchin, have been fully sequenced (Rast and Messier-Solek, 2008). Recently, whole genomic sequencing of four commercial aquaculture species, Atlantic cod (Star *et al.*, 2011), common carp,[146] Nile tilapia,[147] and Pacific oyster,[148] has been completed while genome projects for Atlantic salmon (Davidson *et al.*, 2010), catfish (Lu *et al.*, 2011), rainbow trout, flatfish, and sea bass are underway (Wenne *et al.*, 2007).

For species that lack whole genome sequences, EST data serve as an important resource for gene discovery as well as for the construction of microarrays (Oleksiak, 2010). Thousands of ESTs are available for some species. For instance, currently, there are 498,212 ESTs for Atlantic salmon, 354,488 for

[146]http://english.big.cas.cn/ns/es/201105/t20110509_69516.html.

[147]http://cichlid.umd.edu/cichlidlabs/kocherlab/bouillabase.html.

[148]http://www.pcsga.net/2010/08/chinese-experts-release-worlds-first-oyster-genome-map-2/.

channel catfish, and 287,967 for rainbow trout.[149] Full-length cDNAs, informative tools for functional and structural genome studies, have been obtained for a few aquaculture species, including salmonids (Leong *et al.*, 2010) and catfish (Chen *et al.*, 2010).

The accelerated ability to sequence has led to the detection of thousands of SNPs as markers, that can contribute to the construction of high-density linkage maps and subsequent QTL identification, in species such as Atlantic salmon (Moen *et al.*, 2008), Atlantic cod (Hubert *et al.*, 2010), catfish (Liu *et al.*, 2011a), and sea bass (Kuhl *et al.*, 2011). Diagnostic SNP panels have been assembled to discriminate farmed and wild Atlantic salmon and quantify gene flow (Karlsson *et al.*, 2011). Genomic scans with molecular markers have exposed signatures of selection and shed light on adaptive evolution (Gomez-Uchida *et al.*, 2011; Nielsen *et al.*, 2009).

Functional genomic studies with microarrays (both cDNA and oligonucleotide) can be used to uncover molecular mechanisms underlying productive and adaptive traits in fish (Ferraresso *et al.*, 2008, 2010; Krasnov *et al.*, 2011; McAndrew and Napier, 2010). Microarrays have been developed for numerous species including Atlantic salmon, rainbow trout, common carp, European flounder, Atlantic halibut, African cichlid, and channel catfish. Heterologous microarray analyses, that is, cross-species hybridization, can be performed for expression profiling of closely related nonmodel species for which limited sequence information exists (Healy *et al.*, 2010; Renn *et al.*, 2004). Recently, customized multispecies microarrays have been constructed to assay gene regulation across species (Baker *et al.*, 2009; Kassahn, 2008).

Microarrays have been utilized to quantify gene expression in response to altered environmental conditions such as hypoxia and cold (Boswell *et al.*, 2009; Gracey, 2007), to infection by pathogens such as bacteria, viruses, and fungi (Ewart *et al.*, 2008; Peatman *et al.*, 2007; Roberge *et al.*, 2007; Workenhe *et al.*, 2009) as well as to vaccines (Skugor *et al.*, 2009) and inflammatory stimulators (Djordjevic *et al.*, 2009). Microarray data have yielded insight into diverse biological processes such as nutrition (Taggart *et al.*, 2008), reproduction (Cavileer *et al.*, 2009), stress physiology (Aluru and Vijayan, 2009), larval development (Douglas *et al.*, 2008), life history traits (Aubin-Horth *et al.*, 2005), and migration (Miller *et al.*, 2011).

Transcriptomics has also been used to study adaptive divergence among natural populations, for example, between North Sea and Baltic Sea European flounder populations (Larsen *et al.*, 2007) and between dwarf and normal whitefish (St-Cyr *et al.*, 2008). In addition, gene expression analyses

[149]http://www.ncbi.nlm.nih.gov/dbEST/dbEST_summary.html.

have been applied in aquatic toxicology to evaluate the impact of anthropogenic contaminants and for the discovery of robust biomarkers[150] (Bozinovic and Oleksiak, 2011; Denslow *et al.*, 2007; Hook, 2010). However, to differentiate genetic from environmental effects in natural populations in such experiments, it is essential either that the individuals from different populations are kept under common environmental conditions or that they are subjected to the native environmental conditions experienced by the other populations (Larsen *et al.*, 2011).

Transcriptional data from microarrays, together with previously established linkage maps, can be employed to locate candidate genes (Whiteley *et al.*, 2008). NGS is also proving to be a powerful tool for whole transcriptome characterization and profiling, especially for gene discovery (Jeukens *et al.*, 2010; Johansen *et al.*, 2011a; Xiang *et al.*, 2010). A complementary approach, taking into account both transcriptome and marker data, can be more efficient for defining effective conservation and management strategies (Tymchuk *et al.*, 2010).

The potential of comparative genomics has been exploited to map QTLs and identify candidate genes in nonmodel fish species (Li *et al.*, 2011; Sarropoulou and Fernandes, 2011). Proteomic (Forne *et al.*, 2010; Sanchez *et al.*, 2011) and metabolomic (Flores-Valverde *et al.*, 2010; Samuelsson *et al.*, 2011) approaches are also being increasingly applied in fish.

C. Bioinformatics

Bioinformatics tools for fish are not as comprehensive as for livestock species. In order to coordinate and integrate "omic" data for fish, the Bioinformatics tools for Marine and Freshwater Genomics[151] database has been designed and provides mining tools for analyzing DNA, RNA, and protein sequences (Shih *et al.*, 2010). Species-specific repositories for storing and integrating genomic data as well as access to bioinformatic tools exist for Atlantic salmon,[152] rainbow trout,[153] catfish,[154] tilapia,[155] sea bass,[156] sea bream,[157] and Atlantic cod.[158] A genome-wide fish metabolic network model (MetaFishNet)[159] is available for analyzing high-throughput gene expression data (Li *et al.*, 2010b).

[150]A biomarker is a measurable biological quantity that can be linked to either contaminant exposure or effects.

[151]http://bimfg.cs.ntou.edu.tw/.

[152]http://www.cgrasp.org/index.html.

[153]http://www.irisa.fr/stressgenes/.

[154]http://www.catfishgenome.org/cbarbel/.

[155]http://cichlid.umd.edu/cichlidlabs/kocherlab/bouillabase.html.

[156]http://www.bassmap.org/.

[157]http://www.bridgemap.tuc.gr/index.htm.

[158]http://codgene.ca/index.php.

[159]http://metafishnet.appspot.com/.

D. Cryopreservation

In fish, cryopreservation techniques are mostly applied to sperm and protocols are available for over 200 species of finfish and shellfish, with most studies focused on freshwater species such as salmonids, sturgeons, carps, and catfishes (Diwan *et al.*, 2010). Cryopreservation of fish ova and embryos still remains a challenge primarily due to their large size and biochemical composition (Yang and Tiersch, 2009), although cryopreservation of oocytes (Hamaragodlu *et al.*, 2005; Tervit *et al.*, 2005) and isolated embryonic cells (Hiemstra *et al.*, 2006; Routray *et al.*, 2010) has been demonstrated in some species. Larvae of certain invertebrate species have also been successfully cryopreserved (Kang *et al.*, 2009; Paniagua-Chavez and Tiersch, 2001; Wang *et al.*, 2011).

Unlike in species with internal fertilization, fish sperm is immotile in the testis and seminal plasma and, generally, becomes active and increasingly motile after discharge into the aqueous environment (Alavi and Cosson, 2006). The exception is viviparous fish where the sperm is rendered motile after its release into the female genital tract. The activation mode matches the environment where the sperm functions during spawning and postactivation motility is often of short duration. Thus, dilution of sperm in an extender solution that mimics the osmolality of the seminal plasma and inhibits activation of sperm is crucial for cryopreservation (Viveiros and Godinho, 2009; Yang and Tiersch, 2009).

Thawed sperm is usually less effective than fresh sperm for fertilizing eggs, and therefore, a higher concentration of cryopreserved sperm is needed to achieve the same level of fertilization (Kurokura and Oo, 2008; Kwantong and Bart, 2009) although increasing the sperm–egg ratio too much decreases the fertilization rate (Gwo, 2000; Magyary *et al.*, 2000). Once fertilization has occurred, hatching and survival rates do not seem to differ between fresh and cryopreserved sperm (Chereguini *et al.*, 2001; Kwantong and Bart, 2009). Nevertheless, there have been some reports of sperm cryopreservation promoting DNA damage in fish as well as increased embryonic abortion (Cabrita *et al.*, 2010).

Recently, spermatogonial transplantation into the gonads of sexually mature fish has been established as a novel method to preserve and propagate germplasm from fish that are still sexually immature or are incapable of providing sperm for cryopreservation (Lacerda *et al.*, 2008; Majhi *et al.*, 2009). Spermatogonia also exhibit a high level of sexual plasticity and can produce both functional eggs and sperm, which is especially important for species whose eggs cannot be cryopreserved (Okutsu *et al.*, 2007). Cryopreserved spermatogonia have been successfully transplanted into triploid recipients to generate pure donor-derived male and female progeny (Yoshizaki *et al.*, 2011).

In addition to its advantages described earlier, successful cryopreservation of sperm improves the efficiency of hatcheries by providing sperm on demand and permitting out of season breeding. Operational costs can also be significantly reduced by allowing hatcheries to eliminate the need to maintain live males and potentially diverting resources for use with females and larvae (Tiersch *et al.*, 2007). On the other hand, the economics of integrating cryopreservation into existing fish hatcheries has been investigated and found to be significant. Compared to public hatcheries, investment and operating costs have been found to be higher for private hatcheries, due to associated taxes and interest (Caffey and Tiersch, 2000).

Cryogenic sperm banks have been established mainly in Europe and North America but also in some developing countries such as Brazil, India, and Malaysia (Chew *et al.*, 2010; Harvey, 1998, 2000). Commercial-scale cryopreservation of sperm has also been initiated in a few species such as tetraploid Pacific oysters (Dong *et al.*, 2005) and blue catfish (Hu *et al.*, 2011).

Compared with livestock germplasm, cryopreservation of fish germplasm has had little commercial application. Barriers to widespread application include variation in sperm characteristics among species and stocks, technical problems, a lack of standardization in practices (e.g., gamete collection methods, cryoprotectant choice, evaluation of post-thaw sperm quality, etc.), and inconsistencies within the scientific literature (Tiersch, 2008). Further, research has focused mainly on a small number of species with protocols being unrefined or absent for most species.

E. Reproductive biotechnologies

Reproductive technologies are widely and easily used in many aquatic species. External fertilization, good response to hormones or environmental spawning cues, and the production of large numbers of gametes that are easily manipulated make reproductive technologies an effective means of increasing the use of aquatic genetic resources.

However, reproductive dysfunctions have been observed in some cases. Some males may produce a reduced amount of milt (a serious issue for hatcheries where fertilization is carried out artificially with selected sperm), low-quality milt, or highly viscous milt that cannot fertilize eggs effectively due to diminished dispersal in the water. Captive females often exhibit more severe reproductive problems and fail to undergo oocyte maturation, ovulation, or spawning (Mylonas and Zohar, 2007). In such cases, hormonal treatment is needed for ovulation/spermiation and spawning (Zohar and Mylonas, 2001), and to carry out artificial fertilization, particularly for the creation of hybrids that cannot be produced by natural, spontaneous mating.

1. Hormonal treatment

Control of reproduction allows for the supply of eggs and fry any time during the year, that is, independently of the seasons and the natural spawning time of the species. Hormonal treatment is also used to synchronize ovulation for high seasonal fecundity and a predictable seed supply (FAO, 2011d). The timing of spawning can be altered by the manipulation of species-specific environmental factors such as photoperiod and temperature, but these are not well studied for some species or are nearly impossible to simulate in a hatchery. Consequently, the application of exogenous hormones is the only reliable method to induce spawning.

Hormonal therapies are based on the use of luteinizing hormone preparations such as human chorionic gonadotropin and homologous pituitary extracts; gonadotropin-releasing hormone (GnRH) analogues, with the GnRH analogue-delivery system (with controlled release of the hormone for periods from 1 to 5 weeks) being the most extensively employed in aquaculture (Mylonas *et al.*, 2010); or serotonin-like hormones in the case of bivalves (Braley, 1985). GnRH analogue treatments offer several advantages such as less species specificity, that is, generic use, high potency in small amounts, and no risk of disease transmission from the donor fish to recipient broodstocks (Cnaani and Levavi-Sivan, 2009). However, hormonal treatment can affect gamete quality (particularly eggs) depending upon the nature of the treatment, the gonad developmental stage at which it is applied, and the method of delivery (Avery *et al.*, 2004; Miranda *et al.*, 2005). Hormonal treatments have been shown to be effective in a number of cultured fish species, for instance, for enhancing spermiation and synchronization of ovulation in salmonid fishes (Mylonas *et al.*, 2010), and multiple spawning in carps and flatfish (Agulleiro *et al.*, 2006; Routray *et al.*, 2007) as well as in molluscs, such as giant clams, through serotonin-induced spawning (Braley, 1985).

Sexual dimorphism for economically important traits, such as growth rate, time and age of sexual maturation, and carcass composition, is frequently detected in fish species. For example, male tilapia and channel catfish grow faster than females, whereas salmonid females are larger with higher flesh quality than males. Development of monosex populations is thus preferred in commercial aquaculture production systems (FAO, 2008). Culture of monosex stocks is also advantageous for reproductive containment and for minimizing competition or territorial behavior that occurs in mixed sex populations.

Monosex populations can be generated by manual sorting, interspecific hybridization, and hormonal treatment. While manual sorting is labor intensive and wasteful with a low success rate, interspecific hybridization is applicable only to specific combinations of species and mating results are inconsistent (Cnaani and Levavi-Sivan, 2009). Another impediment to the use of interspecific hybridization, especially in developing countries, is the necessity of increased management for keeping the parental species separated and avoiding hybrid contamination.

Unlike livestock, fish species display some plasticity in their sexual developmental processes, which can be exploited for manipulation of sex. In many species, even though sex is genetically established at fertilization, phenotypic differentiation occurs at a later stage with the timing dependent upon the species involved (Dunham, 2004). The phenotypic sex can be altered either by the application of androgens for masculinization or estrogens for feminization during the critical period of sexual differentiation, generally during early embryogenesis in salmonids and posthatching in cichlids and cyprinids (Cnaani and Levavi-Sivan, 2009). Hormones are usually administered through oral or bath treatment, that is, through the diet or by immersion of eggs and fry in hormone solutions.

The direct induction of sex change by hormones may lead to compromised consumer acceptance even though there is no evidence for the presence of hormonal residues after cessation of the treatment (FAO, 2008). An indirect approach through the combination of sex reversal and breeding is likely to have a wider acceptance, although it requires an understanding of the genetic mechanism of sex determination in the given species and may require progeny testing and multigenerational breeding (FAO, 2011d).

Hormonal sex reversal is widely used on a commercial scale in many countries, for example, for the production of all-male Nile tilapia in Israel, the Philippines, Thailand, and the United States of America[160]; all-female rainbow trout in Europe and the United States of America; and all-female silver barbs in Bangladesh and Thailand (Cnaani and Levavi-Sivan, 2009).

F. Chromosome set manipulation

Chromosomal set manipulation in fish includes inducing polyploidy as well as producing gynogens and androgens. Since most fish and shellfish release their eggs into water before fertilization, they can be easily accessed for manipulation of ploidy levels. This technique has been mainly used to create triploids that are advantageous in production and conservation programs.

Triploids occur naturally in both wild and cultured populations and can be easily induced in many commercial fish and shellfish species. Triploidy is generally induced by physical or chemical shock (with physical treatments being the most successful) and protocols are available for over 30 fish and shellfish species (Dunham, 2004), or alternatively by fertilizing normal haploid eggs with diploid sperm from a tetraploid male (Flajshans *et al.*, 2010; Piferrer *et al.*, 2009). Triploid production by pressure or temperature shock is not 100% effective and species-specific protocols are essential to optimize timing, intensity, and duration of treatment in order to obtain the highest triploid yield (Piferrer *et al.*, 2006). Crossing between tetraploids and diploids is the most efficient method to produce 100% triploids (Dong *et al.*, 2005; Zhou *et al.*, 2010a). There are some data indicating that sperm diploidy should be confirmed prior to fertilization, since

[160]http://www.agribusinessweek.com/update-on-tilapia-sex-reversal/.

not all tetraploid males produce diploid sperm (Nam and Kim, 2004). Tetraploid production, however, is difficult and viable, and fertile tetraploids have been obtained only in a few aquaculture species (Yoshikawa *et al.*, 2008).

Triploid fish are usually effectively sterile, which is desired in conservation programs to prevent introgression of escaped individuals from commercial stocks, including nonnative species, into natural populations and hence reduce the genetic impact of farmed escapees. Sterility is also useful to avoid the establishment of introduced exotic species, for example, triploid grass carp for aquatic weed control (Pípalová, 2006). The migration behavior of triploid fish such as salmon has been investigated and these studies have revealed that the return rate of adult triploid fish to the coast and fresh water was substantially diminished compared to the diploids, that is, triploidization reduces the chances of interbreeding as well as the potential for other negative interactions with wild indigenous salmon in their natal areas (Cotter *et al.*, 2000; Wilkins *et al.*, 2001). Similarly, restocking of all female triploid brown trout did not noticeably impact wild brown trout and the triploid fish did not participate in spawning activity.[161] Triploidy has also been utilized to increase sterility in fertile hybrids (Na-Nakorn *et al.*, 2004), although there is evidence of allotriploid fish producing some viable offspring (Castillo *et al.*, 2007).

Since triploid individuals do not devote energy to gamete production, more energy is available for somatic growth. Performance of triploids is species specific, but they often have similar or less growth than diploids as juveniles and, after maturation, growth is enhanced, especially in shellfish (Piferrer *et al.*, 2009). In addition to increased growth (triploids are 17% larger than diploids prior to spawning and more than 30% larger after spawning), triploid Pacific oysters can be marketed throughout the year due to reduced gonadogenesis (Rasmussen and Morrissey, 2007). Sexual maturity in diploid oysters leads to a decrease in the quality of both taste and texture and thus triploids benefit from higher consumer acceptance (Nell, 2002). Triploid rainbow trout females are preferred for commercial production since they have a better carcass yield during the reproductive season and higher body and filet weights (Werner *et al.*, 2008). Triploidy may also skew sex ratios, for example, in certain species of fish and shellfish, toward a higher proportion of females, which are preferred due to their larger sizes (Cal *et al.*, 2006).

In fish, triploid females are completely sterile and partial to total functional sterility has been observed in males, although offspring sired by artificially produced triploid males are not viable (Feindel *et al.*, 2010; Manning *et al.*, 2004). Nevertheless, since triploid males can still develop secondary sexual characteristics, the production of all-female triploid

[161]http://www.environment-agency.gov.uk/static/documents/Research/triploid_trout_p3_1882654.pdf.

populations (by combining triploidy induction with endocrine feminization) is most desirable to reduce any sexual-related disadvantages (Piferrer, 2001). Triploid shellfish exhibit a decrease in gonadal development as opposed to complete sterility. Moreover, in Pacific oysters, reversion to diploidy has been observed in a proportion of triploid individuals. Verification of triploidy is, therefore, crucial to ensure functional sterility in a given species and it has been proposed that at least two full consecutive reproductive cycles should be monitored for this purpose (Piferrer *et al.*, 2006).

Triploidy is commercially applied mainly in developed countries for numerous fish species such as trout, salmon, charr, and carp, as well as shellfish species such as Pacific oyster, scallops, clams, and mussels (Piferrer *et al.*, 2009). This technology has not yet been commercialized for shrimp (Sellars *et al.*, 2010). Practical implementation of triploidy has not been very successful in developing countries due to high costs and the inability of shock treatment to be 100% effective, a serious impediment to large-scale commercial application (FAO, 2011d). Additionally, in many species, triploid induction results in lower early survival rates, reduced performance under adverse environmental conditions, and increased deformities (Maxime, 2008; Piferrer *et al.*, 2009).

Gynogenesis and androgenesis involve uniparental reproduction, that is, contribution of chromosomes is from one parent only. The sperm (in gynogenesis) or egg (in androgenesis) is irradiated to destroy the chromosomes, followed by fusion with an untreated egg or sperm, respectively, to form a haploid embryo and then diploidy is restored by inhibiting the first mitotic division using temperature or pressure shock (Dunham, 2004). Androgenetic and gynogenetic DHs are thus 100% homozygous and have been successfully induced in many fish species, with reported yields of hatched DHs ranging between 1% and 20%, although survival rates are often low and fertility is reduced in females (Komen and Thorgaard, 2007; Pandian and Kirankumar, 2003; Zhang *et al.*, 2011a). Gynogenesis can also be achieved by the suppression of meiotic divisions (meiosis I in fish and either meiosis I or II in molluscs), but in this case, there are regions of heterozygosity due to recombination.

Gynogens and androgens can be used to study the sex determination mechanism in fish. For example, in species with the XX–XY system, all gynogens are expected to be XX and, therefore, female while androgens should segregate as XX and YY. However, in some cases, deviations from expected progeny sex ratios have been observed due to autosomal and/or environmental factors (Devlin and Nagahama, 2002; Ezaz *et al.*, 2004). Gynogenesis and androgenesis can also be employed to create monosex populations, but due to high mortality of the manipulated eggs, this has not been feasible on a commercial scale. Combined with sex-reversal technology (for instance, crossing XX gynogenetic progeny with sex reversed XX males), large-scale quantities of monosex fish can be produced but this is not yet routine in commercial hatcheries (Beaumont *et al.*, 2010).

A second cycle of gynogenesis or androgenesis can give rise to isogenic clonal lines, which are useful for genetic mapping and QTL analysis. Clonal lines have been successfully established in several commercially important species such as rainbow trout, common carp, amago salmon, and Nile tilapia (Komen and Thorgaard, 2007). Additionally, androgenesis can be applied to recover extinct or endangered species from stored populations of cryopreserved sperm (Babiak *et al.*, 2002; Yasui *et al.*, 2010). Gynogenesis has also been suggested as a tool for accelerating the elimination of recessive deleterious genes and to produce inbred fish with desired traits such as higher growth rates and disease resistance (FAO, 2011d).

G. Biotechnologies for disease diagnosis and prevention

Fish are often reared in open systems where they are exposed to a wide variety of environmental pathogens. Additionally, the horizontal spread of pathogens is facilitated through water and when large numbers of fish are reared within ponds/cages/pens, the increased stress levels make them more susceptible to infection (Adams and Thompson, 2011). The potential effect of diseases spreading from cultured to wild fish and vice versa is another major concern (Johansen *et al.*, 2011b). As aquaculture production intensifies, the likelihood of diseases occurring also increases, which may pose a serious threat to the sustainability and viability of aquaculture as well as economically important wild fish populations.

Akin to the TADs in the livestock sector, transboundary aquatic animal pathogens/diseases (TAAPs/TAADs) are highly contagious and severely impact aquaculture production and trade. Socioeconomic impact studies, carried out in a few countries in Asia and Latin America, have shown that, in addition to a decline in production and sales, TAADs result in export losses, unemployment, and lost consumer confidence (Bondad-Reantaso *et al.*, 2005). Surveys conducted in 16 Asian countries further demonstrate that annual losses due to disease in the region amount to more than USD 3 billion.[162]

The OIE International Aquatic Animal Health Code currently lists 24 diseases (nine for finfish, seven for molluscs, and eight for crustaceans) that require immediate notification in the case of the first occurrence or reoccurrence of a disease in a country or zone of a country that was previously considered to be free of that particular disease, if the disease has occurred in a new host species, if the disease has occurred with a new pathogen strain or in a new disease manifestation, or if the disease has a newly recognized zoonotic potential (OIE, 2010c). In addition to the OIE-listed diseases, the NACA/FAO Quarterly Aquatic Animal Disease Report lists four finfish, three mollusc, and three crustacean diseases as prevalent or exotic in the Asia-Pacific region (NACA/FAO, 2011).

[162]http://www.oie.int/for-the-media/editorials/detail/article/the-role-of-the-oie-in-aquatic-animal-diseases/.

1. Diagnostics

Traditional methods for disease diagnosis tend to be slow and labor intensive. Biotechnological (immunological and molecular) methods, on the other hand, provide powerful tools for rapid and accurate detection and identification of pathogens but can be costly. Often, a combination of traditional and biotechnological methods is needed for definitive disease diagnosis.

Immunological techniques routinely used for the detection of bacterial and viral fish pathogens include direct and indirect fluorescent antibody tests and ELISA, and currently, many commercial antibodies and kits are available (Adams and Thompson, 2008). ELISA is utilized to screen and/or confirm the diagnosis of numerous important pathogens causing diseases such as epizootic hematopoietic necrosis and infectious hematopoietic necrosis in finfish (OIE, 2010d). However, the low sensitivity of antibody-based methods limits their use in environmental samples.

Another limitation is that molluscs and crustaceans do not produce antibodies in response to infection (FAO, 2011d). Nevertheless, in crustaceans, it has been possible to apply ELISA for diagnostics using mouse or rabbit antibodies generated against pathogens purified from infected hosts. For example, diagnostic protocols are available for white spot syndrome virus (WSSV) and *Macrobrachium rosenbergii* nodavirus (OIE, 2010d).

PCR-based diagnostic techniques have been successfully used for the detection of bacteria, viruses, fungi, and parasites in finfish and shellfish (Cunningham, 2002; FAO, 2011d; Poulos *et al.*, 2006). PCR and RT-PCR kits are available and commonly used in the crustacean aquaculture industry for pathogens such as WSSV, Taura syndrome virus and infectious hypodermal and hematopoietic necrosis virus (OIE, 2010d). PCR methodologies have also been extensively used to detect viruses of shrimp broodstock, larvae, and postlarvae in developing countries (FAO, 2011d; Peinado-Guevara and López-Meyer, 2006). For example, in India, the use of PCR-screened postlarvae, together with improved farm management practices, was incorporated into a project for addressing disease and environmental problems in the shrimp industry. Adoption of better management practices led to improved profits, reduced disease incidence, and decreased chemical and antibiotic use (Umesh *et al.*, 2010).

In finfish, PCR protocols for confirmatory testing and diagnosis have been developed for viruses such as epizootic hematopoietic necrosis virus, infectious hematopoietic necrosis virus (IHNV), and koi herpesvirus (KHV); bacteria such as *Vibrio* spp., *Aeromonas* spp., or *Streptococcus* spp.; as well as for parasites such as the ectoparasite *Gyrodactylus salaris* and the oomycete *Aphanomyces invadans* (OIE, 2010d). PCR has also been applied to isolate and identify molluscan disease agents such as *Bonamia ostreae*, *B. exitiosa*, *Marteilia refringens*, *Perkinsus marinus* and *P. olseni*, while real-time PCR is employed for abalone herpes-like virus (OIE, 2010d). Emerging technologies that show promise for diagnostics in fish species include LAMP PCR, which is faster and more sensitive than conventional PCR (Adams and Thompson, 2011; Savan *et al.*, 2005) and DNA microarrays that allow multiplexing for different pathogens (Altinok and Kurt, 2003; Kostic *et al.*, 2008).

The OIE *Manual of Diagnostic Tests for Aquatic Animals* (OIE, 2010d) recommends molecular methods for either direct detection of the pathogen in diseased fish or the confirmatory identification of the pathogen isolated using the traditional method, but not as screening methods for health certification in international trade. FAO and NACA have also developed regional technical guidelines, including for disease diagnosis, for minimizing the risks of disease due to transboundary movement of live aquatic animals in Asia (FAO/NACA, 2000). These guidelines take into account the technical and financial constraints faced by developing countries and accordingly recommend a three-level process based on the existing national diagnostic capacity (FAO, 2011d).

2. Vaccines

Since crustaceans and molluscs do not possess an adaptive or specific immune system, vaccines have been used as prophylatics mainly in finfish and have had a significant impact in reducing disease risk. Vaccination has been applied in the commercial aquaculture of many species, including Atlantic salmon, rainbow trout, sea bass, sea bream, tilapia, turbot, channel catfish, and yellowtail (Håstein *et al.*, 2005). The preponderance of vaccines employed to date have been the traditional inactivated bacterial vaccines. A few inactivated virus vaccines also exist, but no vaccines against fish parasites are commercially available. Despite the absence of a true adaptive immune system, vaccination can be used as an intervention strategy in crustaceans (Pereira *et al.*, 2009) and recombinant subunit vaccines to control WSSV in shrimp are being developed (Vaseeharan *et al.*, 2006; Witteveldt *et al.*, 2004, 2005).

So far, only three attenuated live vaccines have been introduced into the market. Of these, two are live bacterial vaccines, *Edwardsiella ictaluri* that has been licensed for use in salmonids in Chile and North America (Sommerset *et al.*, 2005) and *Flexibacter columnarae* for channel catfish in the United States of America, while the third is a KHV vaccine in Israel (FAO, 2011d). However, environmental and regulatory concerns have been expressed regarding possible reversion to virulence of the attenuated vaccines and the possibility that they might not be nonpathogenic to other relevant species in the wild.

The only DNA vaccine licensed for use in fish is against IHNV in Canada (Salonius *et al.*, 2007). Various DNA vaccines have been tested in experimental challenge trials, with the ones against salmonid rhabdoviruses showing the highest efficacy (Kurath, 2008). DNA vaccines have also been tested for bacterial fish pathogens while vaccines against eukaryotic parasites are still at an early stage of development. A recombinant subunit vaccine against infectious pancreatic necrosis virus is commercially available (Adams *et al.*, 2008) and recently formulations that elicit a strong protective effect against piscirickettsiosis have been developed (Kuzyk *et al.*, 2001; Wilhelm *et al.*, 2006).

Besides contributing to the enhanced health and welfare of cultured fish, in combination with best management practices, vaccination has additional potential benefits such as the ability to grow vaccinated fish at higher densities for improved efficiency, increased appetite and growth in vaccinated fish, and reduced environmental impact due to limited use of antibiotics (Toranzo *et al.*, 2009). For example in Norway, the annual use of antimicrobial agents in farmed salmon declined by 98% over a period of 17 years, due to the introduction of effective vaccines together with improved health management (FAO/OIE/WHO, 2006). In 2007, the average use of antibiotics in the Chilean salmon industry was 1400 times greater than in Norway, with 1.17 kg therapeutant/metric ton produced compared to 0.0008 kg therapeutant/metric ton in Norway (Burridge *et al.*, 2010).

H. Molecular marker-assisted selection

The majority of the genetic improvements in aquaculture species to date have been through conventional breeding methods such as selection, cross-breeding, and hybridization (Hulata, 2001). For traits that are difficult and/or expensive to measure and that are sex limited, expressed after a certain age or require killing the fish, for example, reproductive traits and flesh quality, MAS can substantially increase the accuracy of selection in breeding programs. However, it cannot further reduce the generation interval in fish to less than the age at which sexual maturity occurs.

MAS is only beginning to be applied in aquatic species. Documented cases include QTLs for resistance to infectious pancreatic necrosis in Atlantic salmon (Houston *et al.*, 2008; Moen *et al.*, 2009), for resistance to lymphocystis disease in Japanese flounder (Fuji *et al.*, 2007), and for resistance to pasteurellosis in gilthead sea bream (Massault *et al.*, 2011) and European seabass,[163] that have been incorporated into breeding programs. Nevertheless, before applying this technology on a larger scale and to increase the likelihood of successful MAS, it is necessary to develop high-density linkage maps to speed up the discovery rate of QTLs, especially in species that have less-developed genomic resources.

For species where dense marker maps and high-throughput genotyping have become increasingly available, GS is an attractive option. Simulation studies, carried out to evaluate the potential of GS in aquaculture, have shown that it can yield high genetic gain, accuracy of selection, and very low rates of

[163]http://www.aquagenomics.es/.

inbreeding (Sonesson and Meuwissen, 2009). However, it generates extra costs since results have indicated that continuous phenotypic and genotypic testing is fundamental to maintain the accuracy of the genome-wide breeding values over generations.

I. Transgenesis

Several transgenic aquatic species have been developed (Dunham, 2003) but have not yet been approved for commercial release as food (FAO, 2011d). Approval is, however, reportedly expected in the United States of America for GM Atlantic salmon that grows twice as fast as its wild counterparts.[164]

VII. CURRENT STATUS OF BIOTECHNOLOGIES FOR THE MANAGEMENT OF MICROBIAL GENETIC RESOURCES

Microorganisms are highly diverse and range in effect from harmful to beneficial. Some are responsible for plant and animal diseases, while others play an important role in the improvement of agriculture and food production systems and are multifunctional. However, very few microorganisms associated with food production ecosystems are domesticated. Microorganisms are also subject to relatively rapid evolution as they adapt and change in response to the environment quickly and have short generation times and large populations. The World Data Center for Micro-organisms has records of 596 microbial culture collections held in 68 countries, including many developing countries.[165] Yet, this is just a fraction of the world's microbial diversity and the majority of microbes as well as their ecophysiological roles remain unknown. As detailed below, biotechnology applications have added much to the study of microbial diversity as well as facilitated their utilization in the food industry, plant growth promotion, livestock and fish health, and degradation of pollutants.

A. Molecular markers

The use of molecular markers is especially important in microbes since an estimated 99% of all microorganisms cannot be isolated and thus cannot be characterized based on physiological or biochemical features (Muyzer, 1999). Markers based on the 16S ribosomal RNA (rRNA) gene are most commonly utilized to explore microbial diversity. However, the 16S rRNA gene may occur

[164]http://www.nature.com/news/2010/100914/full/467259a.html.
[165]http://www.wfcc.info/ccinfo/statistics/.

in multiple copies per genome; therefore, markers based on single-copy genes such as the one encoding the RNA polymerase β subunit, *rpoB*, provide better resolution, especially at the subspecies level (Case *et al.*, 2007).

Following amplification by PCR, polymorphisms can be detected by RFLP (PCR-RFLP) or by other methods such as terminal-RFLP (T-RFLP), temperature gradient gel electrophoresis (TGGE), or denaturing gradient gel electrophoresis (DGGE) (Van Elsas and Boersma, 2011). Alternatively, the PCR product can be cloned and sequenced to inventory microbial diversity and gain insights into the phylogeny of species (Le Calvez *et al.*, 2009; Pereira *et al.*, 2006; Porwal *et al.*, 2009).

PCR-RFLP has been utilized to characterize root-nodule bacteria associated with leguminous plants (Hoque *et al.*, 2011) as well as bacteria from the gastrointestinal system of fish (Jensen *et al.*, 2002). T-RFLP is based on the detection of restriction digested fluorescently end-labeled amplified PCR fragments and has been used to study the diversity of complex microbial communities, for example, to evaluate the diversity of bacterial and fungal communities from manure composts at different stages of composting (Tiquia, 2005), to monitor changes in estuarine bacterial communities (Wu *et al.*, 2004), and to identify and discriminate between pseudomonads groups that are involved in the protection of plant roots against soil-borne pathogens (Von Felten *et al.*, 2011).

DGGE and TGGE are types of electrophoresis (applying chemical and temperature gradients, respectively, to denature the sample) that separate similar-sized DNA fragments. DGGE is the most widely used technique for microbial community analyses, for example, to investigate the spatial distribution of soil microbial communities in desert landscapes (Ben-David *et al.*, 2011); compare the phylogenetic diversities of bacteria from aquatic plants (Crump and Koch, 2008); study the dynamics of the bacterial community structure in a wastewater treatment plant (Ding *et al.*, 2011); examine the impact of chlorophenols on the soil bacterial and fungal community composition (Caliz *et al.*, 2011); and explore the diversity and dynamics of bacterial populations during fermentation processes (Jung *et al.*, 2011; Madoroba *et al.*, 2011). TGGE has been employed to deduce the microbial community variation in response to soil type, plant type, or plant development (Wieland *et al.*, 2001) as well as to analyze the bacterial community composition in stream and river water (Beier *et al.*, 2008).

While DGGE and TGGE are useful for characterization at the community level, RFLPs, RAPDs, and AFLPs are more informative for intraspecific diversity and strain-level identification (Gao and Tao, 2012; Maukonen *et al.*, 2003). For example, RAPDs have been employed to detect genetic diversity in different isolates of the fish pathogen *Aeromonas hydrophila* that are phenotypically homogenous (Thomas *et al.*, 2009) and in marine cyanobacterial strains

(Kumari *et al.*, 2009). RFLPs have been used to assess rumen microbial diversity (Deng *et al.*, 2008), and AFLPs have been utilized to investigate the diversity of rhizobia associated with soybean (Wu *et al.*, 2011) and for determining taxonomic and phylogenetic relationships between strains of the nitrogen fixing, filamentous bacteria *Frankia* (Bautista *et al.*, 2011). The Joint FAO/IAEA Division has helped in building the capacities of developing countries to use techniques such as RFLP, DGGE, and TGGE to evaluate rumen microbial diversity.[166]

Microsatellites have been applied to distinguish between fungal species (Lee and Moorman, 2008) as well as to characterize intraspecific diversity in oomycytes (Vargas *et al.*, 2009). Subsequent to whole genome sequencing of many microorganisms, SNPs have been discovered that are aiding in the phylogenetic characterization of bacterial and viral isolates (Gardner and Slezak, 2010) and in elucidating the population structure in fungi (Abbott *et al.*, 2010; Broders *et al.*, 2011).

With reference to pathogenic microorganisms, a multinational consortium, Quarantine Barcoding of Life,[167] is developing a DNA barcode identification database for use in plant health diagnostics. The consortium aims to generate barcode sequences for quarantine species and closely related species, to develop generic diagnostic tools based on these barcode sequences, and to develop methodologies to enable the establishment of DNA banks and access to digital specimens (Bonants *et al.*, 2010).

B. "Omics"

Genomes of microbes are typically small in size and hence easy to sequence. Consequently, over 1000 microbial genomes have been sequenced and several more are nearing completion (Kumar *et al.*, 2011). The availability of genome sequences offers new tools to better characterize candidate starter strains for fermentation processes (Kleerebezem and de Vos, 2011; Torriani *et al.*, 2011). Comparative genome hybridization analysis with multistrain microarrays can be used to investigate intraspecific diversity and to correlate this with industrially relevant phenotypic diversity (Borneman *et al.*, 2010; Siezen *et al.*, 2011). Genome mining of bacteria and fungi has unveiled the secondary metabolite product potential of these microorganisms (Winter *et al.*, 2011).

Genome sequencing of rumen microbes is being employed to identify targets for methane mitigation technologies (Attwood *et al.*, 2011; Buddle *et al.*, 2011) as well as to understand the functioning of microbes within the rumen and

[166]http://www-naweb.iaea.org/nafa/aph/crp/aph-molecular-techniques.html.
[167]http://www.qbol.wur.nl/UK/about-qbol/.

their impact on ruminant health and performance (Suen *et al.*, 2011). Whole genome sequencing of the host together with its symbiont/pathogen, for example, *P. trichocarpa* and its fungal symbiont *Laccaria bicolor*, is a powerful tool to gain new perspectives on the coevolution of their genomes (Medina and Sachs, 2010).

Advances in microbial transcriptomics have shed additional light on plant–microbe interactions (Bonfante and Genre, 2010), rumen microbial processes (Dodd *et al.*, 2010), and livestock host–pathogen relationships (Bannantine and Talaat, 2010). Transcriptomic analyses of food-borne bacterial pathogens in response to specific stresses (associated with food hygiene, processing, and preservation measures) have been carried out to enable more effective control strategies to be devised (King *et al.*, 2010; Soni *et al.*, 2011; Zhang *et al.*, 2011b).

A growing number of proteomic approaches are focusing on both symbiotic and pathogenic plant–microbe interactions to unravel the molecular communication between plants and microbes, including the identification of proteins involved in suppression of plant defense responses by symbiotic microbes and of virulence factors responsible for microbial pathogenicity (González-Fernández *et al.*, 2010; Mathesius, 2009; Quirino *et al.*, 2010). Proteomic analyses of livestock and fish pathogens have been initiated to uncover virulence mechanisms and to increase the gamut of antigens available for the formulation of diagnostics and vaccines (Dumpala *et al.*, 2010; Menegatti *et al.*, 2010; Pinto *et al.*, 2009).

Metagenomic approaches are becoming more feasible and cost–effective and are increasingly used to characterize microbial genetic diversity. For example, they are providing insight into bacterial phylogeny (Rosen *et al.*, 2009), into the ecology of soil microbes to optimize agricultural production and to identify novel biomolecules (Ghazanfar *et al.*, 2010; Mocali and Benedetti, 2010), as well as into rumen microbial function to discover biomass-degrading enzymes (Ghazanfar and Azim, 2009; Hess *et al.*, 2011). Nonetheless, efforts should continue to be made to culture microorganisms so that the sequence information is used in conjunction with isolation and cultivation to determine the ecophysiological roles of microbes (Yamada and Sekiguchi, 2009).

An integrated strategy, utilizing multiple "omic" technologies, is particularly useful to fully understand the functioning of microbial systems. Several combined transcriptomic and proteomic analyses have been undertaken, to complement each other thus avoiding the detection bias from each technology, for cross-validation purposes and to reveal novel insights into metabolic and regulatory processes (Zhang *et al.*, 2010a). Similarly, integrated transcriptomics and metabolomics analysis can be a robust tool to elucidate the molecular machinery underlying physiological processes (Yang *et al.*, 2009b).

C. Bioinformatics

Fungal genome and proteome databases include the Fungal Genome Initiative of The Broad Institute,[168] the Sanger Institute fungal sequencing effort,[169] the *Saccharomyces* Genome Database,[170] the *Fusarium graminearum* Protein-Protein Interaction Database,[171] and *Magnaporthe grisea* Protein–Protein Interaction Database[172] (González-Fernández *et al.*, 2010). Several bacterial databases exist, among them the *Escherichia coli* specific EchoBASE[173] and the annotation of microbial genome sequences[174] system that incorporates tools for integrated high-throughput genome annotation and protein function prediction for completed and draft bacterial genomes (Kumar *et al.*, 2011).

The G-InforBIO Database[175] (Tanaka *et al.*, 2006) and the Microbial Genome Database (MBGD)[176] facilitate comparative genomic analyses of microbial sequences. The Pathogen–Host Interaction Database catalogs experimentally verified pathogenicity, virulence, and effector genes from microbial pathogens that infect plant and animal hosts.[177] Bioinformatic tools and databases for metagenomics include the Genomes OnLine Database,[178] the MEtaGenome ANalyzer,[179] StrainInfo,[180] and UniFrac[181] (Mocali and Benedetti, 2010; Singh *et al.*, 2009).

D. Cryopreservation

According to the World Federation for Culture Collections guidelines for the establishment and operation of collections of cultures of microorganisms, in order to minimize the probability of strains being lost, each strain should be maintained by at least two different procedures, and at least one of these should be by lyophilization (freeze-drying) or cryopreservation (WFCC, 2010). Cryopreserving microorganisms circumvents the need for repeated subculturing,

[168]http://www.broadinstitute.org/scientific-community/science/projects/fungal-genome-initiative/fungal-genome-initiative.

[169]http://www.sanger.ac.uk/Projects/Fungi/.

[170]http://www.yeastgenome.org/.

[171]http://csb.shu.edu.cn/fppi/.

[172]http://bioinformatics.cau.edu.cn/cgi-bin/zzd-cgi/ppi/mpid.pl.

[173]http://www.york.ac.uk/res/thomas/howtouse.htm.

[174]http://www.bhsai.org/ages.html.

[175]http://www.wfcc.info/inforbio/G-InforBIO/download.html.

[176]http://mbgd.genome.ad.jp/.

[177]http://www.phi-base.org/.

[178]http://www.genomesonline.org/.

[179]http://www.megan-db.org/megan-db/home/.

[180]http://www.straininfo.net/.

[181]http://bmf.colorado.edu/unifrac/.

which is time-consuming and prone to contamination. Additionally, it prevents the genetic and physiological changes that could be associated with the long-term culture of actively growing organisms (Day *et al.*, 2008). Cryopreservation can be used to preserve microbes that cannot be cultured or those that require growth on their host. It is also the most suitable long-term method for maintaining the stability of secondary metabolite production (Ryan *et al.*, 2003).

Numerous factors influence the effectiveness of cryopreservation of microorganisms, such as species, cell type, age, growth phase, osmolarity, and aeration, and composition of the freezing medium. For example, microorganisms, especially bacteria, that are grown under aerated conditions and are harvested in late log or early stationary phase are more resilient to freezing stress (Kerrigan, 2007). Furthermore, the presence of cryoprotective additives such as sulfoxides, alcohols and derivates, saccharides and polysaccharides, and glycoproteins in the freezing medium can enhance survival rates considerably (Hubalek, 2003). Cooling rates can also vary considerably, with the majority of fungi being successfully preserved at a cooling rate of $-1\ °C\ min^{-1}$, while some require a rate of $-10\ °C\ min^{-1}$ (Smith and Ryan, 2008).

Many microorganisms have been successfully cryopreserved, including over 4000 species of fungi, belonging to over 700 genera (Smith *et al.*, 2008). Cryopreserved microorganisms usually exhibit high survival rates (often reaching levels of 100%) and good strain stability.[182] For instance, viability of lyophilized yeast is typically between 1% and 30% as opposed to greater than 30% for cryopreserved yeast (Smith *et al.*, 2008). However, selected bacterial species are sensitive to cryopreservation, especially in terms of viability and stability of antigenic, molecular and biochemical properties, and improved protocols need to be developed (Paoli, 2005).

E. Pathogen detection in food

Food-borne diseases are a major concern for public health and cause high levels of morbidity and mortality,[183] with *Campylobacter*, *Salmonella*, *Listeria monocytogenes*, and *E. coli* O157:H7 generally being responsible for the majority of food-borne outbreaks (Velusamy *et al.*, 2010). Therefore, the rapid and reliable detection of microbial pathogens in food is essential for ensuring food safety and quality. Microbial analysis also plays a valuable role in assessing the performance of management strategies based on the Hazard Analysis and Critical Control Point (FAO, 1998, 2006).

[182]http://www.wfcc.info/tis/info3.pdf and http://www.wfcc.info/tis/info4.pdf.
[183]http://www.who.int/foodsafety/foodborne_disease/en/.

Conventional methods, such as culture and colony-counting methods, are sensitive and inexpensive but often time-consuming and labor intensive. To overcome these limitations, a diverse array of biotechnological methods has been developed and is available for the improved detection of food-borne pathogens (Jasson *et al.*, 2010; Mandal *et al.*, 2011). ELISA is the most prevalent antibody assay for the detection of bacterial pathogens and contaminants in food, and numerous ELISA tests are commercially available as robotized automated systems (Jasson *et al.*, 2010). Commercial ELISA kits are also available for detecting mycotoxins in food and animal feed (Schmale and Munkvold, 2009).

PCR-based methods are used in the identification of a wide range of pathogens such as *L. monocytogenes* (Zunabovic *et al.*, 2011), *Staphylococcus aureus* (Riyaz-Ul-Hassan *et al.*, 2008), *Salmonella* (Stark and Made, 2007), and *E. coli* O157:H7 (Perry *et al.*, 2007) as well as for the simultaneous detection of pathogens (Kim *et al.*, 2007; Mukhopadhyay and Mukhopadhyay, 2007; Zhang *et al.*, 2009). Quantitative real-time PCR has also been applied for the accurate quantification of pathogens in food samples (Alarcon *et al.*, 2006; Malorny *et al.*, 2008; Rantsiou *et al.*, 2008), and several PCR and real-time PCR-based diagnostic kits have been commercialized (Glynn *et al.*, 2006).

In recent years, the potential of biosensors in food-borne pathogen detection has been investigated (Gehring, 2011). A biosensor is an analytical device that converts a biological signal into an electrical one. It uses a bioreceptor, that is, an immobilized biologically related agent (such as an nucleic acid, enzyme, antibiotic, organelle, or whole cell), to detect or measure a chemical compound, and a transducer to convert the recognition event into an electric signal (FAO, 2001; Velusamy *et al.*, 2010). Currently, biosensors are being optimized and validated for the routine monitoring and/or identification of bacteria, including field-based pathogen detection (Setterington and Alocilja, 2011; Zordan *et al.*, 2011).

F. Food preservation and production of food and feed ingredients

Fermentation is a cost–effective food preservation process that can also enhance the flavor, aroma, and texture of food, enrich its nutritional quality and digestibility, detoxify contaminated food, and decrease cooking time and fuel requirements (Liu *et al.*, 2011b). In many developing countries, fermented foods serve as important dietary constituents and are produced primarily at the household and village level. As such, the majority of small-scale fermentations are based on spontaneous processes, resulting from the activities of a variety of microorganisms associated with the raw food material and the environment. The majority of fermented foods in Africa are produced by spontaneous fermentation, for example, Cingwada (fermented cassava) in East and Central Africa, Kenkey (fermented maize) in Ghana, and Owoh (fermented cotton seed) in West Africa (FAO, 2011e).

However, limitations include enhanced lag phase of microbial growth associated with contamination of competing microorganisms, that is, higher probability of spoilage, variable product quality, and lower product yield (Holzapfel, 2002).

Starter cultures are preparations of live microorganisms that are added to initiate and/or accelerate fermentation processes (FAO, 2011e). The starter culture may be obtained through the practice of back slopping (addition of a sample from a previous successful fermentation batch) or may be a "defined starter culture" consisting of single or multiple strains normally produced by pure culture maintenance and propagation under aseptic conditions (FAO, 2011e). Examples of fermented foods produced using a back-slopping process include fermented cereals and grains in Africa and fermented fish sauces and vegetables in Asia (FAO, 2011e). Strains selected for defined starter cultures should possess several desirable metabolic traits, lack toxicogenic activity, and also be suitable for large-scale production (Gänzle, 2009). Defined starter cultures allow process standardization together with lowered health risks and often incorporate adjunct cultures to inhibit pathogenic or food-spoilage organisms and to improve product quality (Mendoza *et al.*, 2011; Settanni and Moschetti, 2010).

Lactic acid bacteria are the predominant microorganisms in a preponderance of food fermentations. They convert carbohydrates either to lactic acid alone or carbon dioxide and ethanol in addition to lactic acid and are responsible for many products such as fermented sausage, all fermented milks, pickled vegetables, and sour dough bread (Flores and Toldra, 2011; Liu *et al.*, 2011b; Steinkraus, 2002). Acetic acid bacteria are important in the food industry due to their ability to oxidize sugars and alcohols into organic acids and are used in the production of vinegar and in cocoa and coffee fermentations (Sengun and Karabiyikli, 2011). A third group of bacteria, belonging to the genus *Bacillus*, hydrolyze proteins to amino acids and peptides and release ammonia. Such alkaline fermentations of plant seeds as well as legumes provide protein-rich condiments especially in Africa and Asia (Parkouda *et al.*, 2009). Yeast fermentations, generally involving *Saccharomyces* species, result in the formation of ethanol and carbon dioxide from sugar and are widely used for the production of leavened bread and fermented beverages such as wines and beers (Sicard and Legras, 2011).

Fermentation that leads to the nutritional fortification of traditional foods can have a profound impact on the diets of people in developing countries that depend largely on one staple, such as cassava, maize, or rice, for subsistence. For example, fermentation of rice to produce tape ketan in Indonesia results in a doubling of protein content and enrichment with lysine, an essential amino acid. Similarly, pulque that is produced by the fermentation of agave juice in Mexico is rich in vitamins like thiamine, riboflavin, niacin, biotin, and pantothenic acid (Steinkraus, 2002).

Essential amino acids, produced by microbial fermentation, are also utilized to supplement grain-based livestock feeds, both to increase productivity and to decrease the excretion of nitrogen from the animals into the environment (FAO, 2011c). Currently, the annual global use of L-lysine, the first limiting amino acid for pigs and the second limiting amino acid after methionine for poultry, is estimated to be 900,000 tons followed by 65,000 tons for L-threonine and 1900 tons for L-tryptophan (Kim, 2010). Feed grade L-valine is marketed in the EU, while L-glutamine, also produced through fermentation processes, is available in South America and selected Asian countries (Kim, 2010). Additionally, exogenous microbial enzymes are increasingly being incorporated in animal feeds. Supplemental phytase, the most extensively used feed enzyme, improves utilization of phosphorus as well as other minerals in pigs and poultry and can reduce phosphorus excretion by up to 50% (Singh *et al.*, 2011b). Phytase has recently also been approved for use in salmonid feed in the EU.[184] Other exogenous enzymes included as feed additives to ameliorate digestion are xylanases, glucanases, proteases, and amylases (FAO, 2011c).

Microbial enzymes, manufactured by fermentation under controlled conditions, are commonly employed in the food processing industry. For example, α-amylases are applied for converting starch into fructose and glucose syrups (Souza and Magalhães, 2010), proteases such as chymosin are used in cheesemaking, pectinases are utilized for extraction, clarification and concentration of fruit juices, and tannases are used for the production of instantaneous tea (Aguilar *et al.*, 2008). Microorganisms are also used to generate volatile flavor chemicals that possess desirable properties such as antimicrobial and antioxidant activities in addition to sensory properties, and more than 100 aroma chemicals are available commercially (Berger, 2009). In recent years, there has been a growing interest in exploiting microbial fermentation processes for the production of bioethanol and biodiesel (Cheng and Timilsina, 2010; Demain, 2009; Ruane *et al.*, 2010; Shi *et al.*, 2011).

G. Biofertilizers

Soils are complex, interactive, and dynamic living systems that contain a variety of microbes. There has been increasing worldwide concern regarding the adverse effects of the indiscriminate use of chemical fertilizers on soil productivity and environmental quality. Based on beneficial plant–microbe interactions, it has been possible to develop microbial inoculants or biofertilizers, that are cost effective and environment friendly, for use in agriculture to improve plant performance. In addition to stimulating plant growth and reducing soil pollution,

[184]http://www.fishnewseu.com/latest-news/world/5553-phytase-product-receives-eu-approval-as-feed-additive-for-salmonids.html.

biofertilizers can improve soil structure and fertility and provide protection against abiotic stresses as well as a broad range of diseases (Pineda *et al.*, 2010; Yang *et al.*, 2009c).

Nitrogen fixing rhizobial inoculants are the most widely used biofertilizers (Mia and Shamsuddin, 2010) and are applied to leguminous crops as well as other commercially important crops such as rice, maize, sugarcane, and wheat, leading to increased grain yield and/or biomass (Bhattacharjee *et al.*, 2008). Often, inoculating a mixture of bacterial isolates produces a synergistic result (Govindarajan *et al.*, 2008), although it is important to determine the compatibility of the strains in the mixture prior to inoculation.

Numerous rhizobial inoculants have been commercialized in developing countries, for example, Biofix in Kenya (Odame, 2002) and Rhizofer in Mexico (FAO, 2011a). In Argentina, Bolivia, Paraguay, and Uruguay, 70% of the soybean crop area is estimated to be inoculated with rhizobia.[185] A study in Thailand showed that the use of rhizobial inoculants, between 1980 and 1993, in soybean, groundnut, and mungbean production, led to estimated accumulated benefits of approximately USD 100, USD 17, and USD 4 million, respectively (Boonkerd, 2002). However, another study in Thailand highlighted the varying performance of inoculants in different locations, highlighting the role of farmers' knowledge and experience in the effective application of biofertilizers (Sonnino *et al.*, 2009).

A variety of free-living cyanobacteria are found naturally in rice fields, with *Anabaena* and *Nostoc* being the most common nitrogen-fixing organisms. Cyanobacterial application in rice fields can reduce the use of urea fertilizer by 25–35% (Hashem, 2001) and has resulted in enhanced grain yield in China, Egypt, India, Japan, the Philippines, and other rice-growing tropical countries (Vaishampayan *et al.*, 2001). Soil-based cyanobacterial biofertilizers can be produced with ease, with the cost of inoculum preparation being one-third that of chemical fertilizers (Sharma *et al.*, 2011). However, inoculation of rice fields with selected strains is not a proven technology. Due to problems in the establishment of inoculated strains and their inconsistent performance in increasing yields, there has been localized use of cyanobacterial biofertilizers as inoculants (Prasanna *et al.*, 2011).

A few cyanobacterial species also form symbiotic associations, for example, *Anabaena azollae* with the water fern *Azolla*. The *Azolla-Anabaena* association has been used as a biofertilizer in Brazil, China, India, Indonesia, Italy, Mexico, the Philippines, Taiwan, Thailand, Viet Nam, and several Western African countries (Carrapiço *et al.*, 2000; Vaishampayan *et al.*, 2001), but in recent years, the area devoted to *Azolla* has decreased due to the technology being labor intensive, the availability of cheap sources of urea and potash as well

[185]http://info.ipni.net/biofertilizers.

as changing agricultural practices and policies (Roger, 2004). Additional constraints limiting its use include difficulties in maintaining inocula throughout the year, phosphorus limitations in soils, availability and control of water supplies, low tolerance to high temperature, and damage by pests (Choudhury and Kennedy, 2004). Regardless, *Azolla* has potential as a multipurpose crop since, in addition to being a nitrogen source for rice crops, it can be employed for reclaiming saline soils, to control weed infestations, to purify wastewater and as an animal feed (Choudhury and Kennedy, 2004; Roger, 2004). For example, a technology called "Azobiofer" has been developed for the production and utilization of *Azolla* for irrigated rice and fish cultivation, which could be profitable for tropical and subtropical rice growing countries (Mian, 2002).

Phosphate solubilizing (P-solubilizing) microorganisms play a key role in the availability of phosphate to plants and include bacteria such as *Bacillus* and *Pseudomonas* and fungi such as *Aspergillus* and *Penicillium*. Bacteria are more effective in phosphorus solubilization since P-solubilizing bacteria (PSB) constitute 1–50% of the natural bacterial population in soil, while P-solubilizing fungi (PSF) are only 0.1–0.5% of the total fungal population (Khan *et al.*, 2009a). Commercial PSB biofertilizers include BioP-Plus (*Bacillus coagulans*)[186] and Sukrish-p[187] while JumpStart (*Penicillium bilaiae*) and PR-70 RELEASE (*P. radicum*) are examples of commercially released PSF biofertilizers (Khan *et al.*, 2010).

Often, a synergistic effect on plant growth has been observed by coinoculation of P-solubilizing bacteria and fungi. For example, a microbial preparation termed Indian Agricultural Research Institute microphos culture was developed in India that contained two efficient PSB (*Pseudomonas striata* and *Bacillus polymyxa*) and three PSF (*Aspergillus awamori*, *A. niger* and *Penicillium digitatum*) (Khan *et al.*, 2010). Coinoculation of P-solubilizing microorganisms with nitrogen fixers such as *Azospirillum* and *Azotobacter* (Yadav *et al.*, 2011b) as well as with arbuscular mycorrhizal fungi (Khan *et al.*, 2009b) has also been shown to have a positive effect on the growth and nutrient uptake of plants.

Mycorrhiza are fungi that form a symbiotic association with plant roots and increase plant uptake of phosphorus and zinc in exchange for carbohydrates. The mycorrhizas that are ecologically and economically relevant for agriculture include ectomycorrhizas (that do not penetrate the cells of the plant) and endomycorrhizas such as arbuscular mycorrhizal fungi (AMF) that penetrate the cells forming clusters/arbuscules and vesicles (Habte, 2000). AMF are the most abundant and widely distributed type and are found in about 95% of land plant species. Although AMF have a very broad host range, species can vary in abundance in response to different hosts and environmental conditions (Van Diepen *et al.*, 2011).

[186]http://www.soo.co.in/biopplus.htm.

[187]http://www.indiamart.com/totalagricare/organic-bio-fertilizer.html.

In addition to improved access to nutrients and water, AMF enhance plant productivity through pest and disease suppression, increased drought tolerance and alleviating the detrimental effect of salinity and alkalinity (Cardarelli *et al.*, 2010). More than 30 companies worldwide produce commercial mycorrhizal fungal inoculum (Schwartz *et al.*, 2006). The application of mycorrhizal fungi has been especially important in forestry where it has contributed to improved tree growth (Ouahmane *et al.*, 2007), reafforestation of degraded soils (Duponnois *et al.*, 2008) as well as to minimizing the negative influence of exotic tree species on soil microbial communities (Kisa *et al.*, 2007). However, there are also some potential concerns regarding their application such as the detrimental effects on host plants and negative impact on the diversity of local fungal and plant communities (Schwartz *et al.*, 2006).

Production of organic fertilizer through composting can also be hastened with the addition of cellulose decomposing fungi. For example, in the Philippines, inoculation of substrates such as rice straw with *Trichoderma* reduced composting time to 21–45 days, depending on the type of plant residues used. Moreover, rice and sugarcane farmers that adopted this technology used less chemical fertilizers and had higher yields and net incomes (Sonnino *et al.*, 2009).

H. Biopesticides

Biopesticides are biological alternatives to chemical pesticides and as such are environmentally sustainable pest management tools. There has been a rising demand for biopesticides in agriculture due to increased application of IPM, lower risk to human health and enhanced safety, and minimal negative impact on the environment. Global percentage of biopesticides has thus steadily grown, and currently, Canada, Mexico, and the United States of America dominate the market, employing 44% of the products sold worldwide, followed by the EU and Oceania with 20% each, Latin America at 10%, and Asia with 6% (Bailey *et al.*, 2010). In contrast, the global market for synthetic pesticides has declined, with a total negative growth of 7.4% from 1990 to 2005 (CropLife International, 2006).

Microbial biopesticides offer many advantages, such as high levels of specificity combined with lower developmental and registration costs than those for conventional pesticides (Chandler *et al.*, 2008). They are thus particularly attractive and affordable options for developing countries. As of 2010, the number of microbial biopesticides registered/available worldwide were 327 in China, 72 in the United States of America, 42 in Brazil, 40 in Canada, 39 in the EU, 20 in South Africa, 20 in Ukraine, 17 in Cuba, 15 in India, 14 in New Zealand, 13 in the Russian Federation, 11 in Australia, 11 in Kenya, 11 in the Republic of Korea, 9 in Argentina, and 8 in the Republic of Moldova (Kabaluk *et al.*, 2010).

Microbial biopesticides have had considerable success in controlling plant pests. In China, the fungus *Beauveria bassiana* has been deployed against the Masson's pine caterpillar, a serious forest defoliator. Over a period of 36 years, various applications of this fungus were able to successfully control the forest pest and dramatically reduce the use of chemical pesticides (Li, 2007). In Brazil, during the growing season of 2007/2008, the fungus *Metarhizium anisopliae* was applied to 250,000 ha in São Paulo alone for control of the sugarcane root spittlebug (Kabaluk *et al.*, 2010), while in 2004/2005, the *Anticarsia gemmatalis* nucleopolyhedrosis virus was used on 2 million hectares to control the velvet bean caterpillar, a key soybean pest (Sósa-Gomez *et al.*, 2008). In India, *Hyblaea puera* nucleoploy-hedrovirus (HpNPV) is being commercialized as a biocontrol agent of the teak defoliator *H. puera*, an important pest of teak plantations causing heavy losses in volume increment (Kumar and Biji, 2009; Wahab, 2009). The proprietary agent Green Muscle®, composed of the spores of the fungus M. *anisopliae* and a mixture of mineral oils, has been instrumental in containing red locust infestations in Africa. An FAO intervention for locust control with Green Muscle® in 2009 prevented an invasion that could have potentially damaged food crops of about 15-million people in Eastern and Southern Africa.[188]

I. Bioremediation

A huge amount of pollutants such as heavy metals, mineral oil, polycyclic aromatic hydrocarbons, aromatic hydrocarbons, and phenols are expelled into the environment every year. In Europe alone, it is estimated that potentially polluting activities have occurred at approximately 3 million sites, of which only about 80,000 sites have been remediated in the past 30 years (Guimarães *et al.*, 2010). The global intensification of aquaculture has also led to escalating amounts of effluents being discharged that contain feed residues, faces, and antibiotics. The resulting organic enrichment causes environmental deterioration, negatively impacts the ecosystem and may lead to an increased presence of pathogenic bacteria as well as contributing to the spread of viruses (Chávez-Crooker and Obreque-Contreras, 2010).

Treatment of contaminated soil and water can be carried out with traditional physical, chemical, and thermal processes, but these are often expensive. Bioremediation is a more efficient, cost–effective, and environment-friendly technology for the clean-up and restoration of contaminated sites. Many species of bacteria (e.g., *Bacillus*, *Haemophilus*, *Mycobaterium*, *Pseudomonas*, and *Rhodococcus*) and fungi (e.g., *Aspergillus*, *Cladosporium*, *Penicillium*, *Stropharia*, and *Trichoderma*) can degrade toxic pollutants, with bacteria being the most active agents of bioremediation (Fernández-Luqueño *et al.*, 2011; Tyagi *et al.*, 2011).

[188]http://www.fao.org/news/story/percent20en/item/21084/icode/en/.

Bioremediation strategies involve either bioaugmentation, that is, inoculation of the contaminated site with enriched microorganisms that degrade the target contaminants, or biostimulation, that is, the addition of nutrients to favor degradation by a given native microbial population. Bioaugmentation usually works best when microorganisms are preselected from a contaminated site and a multistrain consortium is employed (Alisi *et al.*, 2009; Li *et al.*, 2009b). However, it has been observed that the number of introduced microorganisms decrease shortly after addition to a site due to abiotic and biotic stresses, competition for limited resources from indigenous microbes as well as antagonistic interactions (Jacques *et al.*, 2009).

Adding one or more rate-limiting nutrients, by adding organic matter such as manure, compost, wastewater sludge, etc. accelerates the decontamination rate in biostimulation processes (Fernández-Luqueño *et al.*, 2011). A disadvantage of biostimulation compared to bioaugmentation is that its effect is delayed due to the lag between nutrient application and propagation of the microbial population. The decision to implement either or both of these strategies depends upon the circumstances prevailing at the contaminated site and a complementary approach is emerging as promising. Microorganisms can also be utilized to increase the bioavailability of hydrocarbons for subsequent degradation through the synthesis of surface-active compounds or biosurfactants (Banat *et al.*, 2010; Das *et al.*, 2008).

A number of commercial bioaugmentation and biostimulation products are available on the market, for example, DBC-plus™, Biosolve and S-200, although more evaluation is necessary prior to their implementation in diverse environments (Tyagi *et al.*, 2011). On-site bioremediation studies have been carried out in a few countries including Canada (Sanscartier *et al.*, 2009), China (Liu *et al.*, 2010), Japan (Tsutsumi *et al.*, 2000), and Spain (Jimenez *et al.*, 2006). Bioremediation in aquaculture has also been reported, mainly using bacteria in shrimp hatcheries (Kumar *et al.*, 2009; Manju *et al.*, 2009).

J. Probiotics

Worldwide concern about increased antibiotic administration in livestock and fish and the resulting emergence of antibiotic resistance has led to searches for alternative and effective approaches, such as the use of probiotics. Probiotics have also been reported to play an important role in immunological, digestive and respiratory functions in humans. Once ingested, probiotics produce beneficial physiological effects, although the mechanisms by which these effects are exerted are still poorly understood. Proposed modes of action include competitive exclusion of pathogenic bacteria by competing for nutrients, space, and iron; production of antagonistic compounds that inhibit the growth of harmful bacteria; modulation of the host's immune system; and enhanced feed conversion efficiency by providing dietary compounds and enzymes (Farzanfar, 2006; Nayak, 2010b; Soccol *et al.*, 2010; Vine *et al.*, 2006). A successful probiotic should itself be nonpathogenic and nontoxic, benefit the host, have the capacity to be adherent, and colonize the gastrointestinal tract, replicate to high numbers and be stable for long periods under storage as well as in field conditions (Farzanfar, 2006; Soccol *et al.*, 2010).

In humans, several studies on potential health benefits of probiotics have been performed. Studies have been carried out on prevention of relapses of inflammatory bowel diseases and ulcerative colitis (Wohlgemuth *et al.*, 2010), prevention of antibiotic associated diarrhea (Fitton and Thomas, 2009), reduction of the duration of acute diarrhea in children (Henker *et al.*, 2008), reduction of the duration of common cold and flu infections, and lowering of the risk of developing eczema if taken by pregnant women and their infants in early life (Weichselbaum, 2009). Many different food products have been formulated with probiotics and are available commercially, such as fermented milks with high or low viscosity in the EU (Soccol *et al.*, 2010); yogurt-covered raisins, nutrient bars, chocolate bars, and tablets in the United States of America; ice cream in Latin America; and yogurt in the Republic of Korea (Burgain *et al.*, 2011). With regards to the assessment of the efficacy and safety of probiotics, guidelines have been developed jointly by FAO and WHO that provide a methodology for use in the evaluation of probiotics and define the criteria and specific levels of scientific evidence needed to make health claims for probiotic foods (FAO/WHO, 2006).

The use of probiotiocs in aquaculture is relatively new, but many studies have been undertaken regarding their application in finfish (Balcazar *et al.*, 2006), crustaceans (Farzanfar, 2006), molluscs (Prado *et al.*, 2010) as well as larvae (Vine *et al.*, 2006). Probiotics are dispensed through feed and/or as a water additive, with supplementation through feed being a better method for successful colonization and establishment in the gut (Nayak, 2010b). Numerous health benefits have been attributed to probiotics in fish, for example improved growth (Boonthai *et al.*, 2011; Lara-Flores *et al.*, 2003), stimulation of the immune system (Picchietti *et al.*, 2009; Zhou *et al.*, 2010b), enhanced survival rate (Faramarzi *et al.*, 2011; Hjelm *et al.*, 2004), facilitation of feed utilization and digestion (Manju *et al.*, 2011; Sun *et al.*, 2011) as well as control of pathogenic microorganisms and disease resistance (Balcazar *et al.*, 2009; Vendrell *et al.*, 2008).

The application of probiotics has been well documented in livestock. In poultry, administration of probiotics has been shown to reduce chick colonization and invasion by *Salmonella* (Higgins *et al.*, 2008; Vila *et al.*, 2009), decrease mortality (Timmerman *et al.*, 2006), increase body weight (Mountzouris *et al.*, 2007), and improve egg production and quality (Panda *et al.*, 2008). In pigs, probiotic inclusion in feeds has been demonstrated to decrease the pathogen load (Collado *et al.*, 2007; Taras *et al.*, 2006), ameliorate gastrointestinal disease symptoms (Zhang *et al.*, 2010b), and improve weight gain (Konstantinov *et al.*, 2008). In ruminants, supplementation with probiotics has been reported to result in enhanced milk yield (Desnoyers *et al.*, 2009), improved growth performance (Frizzo *et al.*, 2010), increased dry matter intake (Jouany, 2006), reduced incidence of diarrhea (Von Buenau *et al.*, 2005), increased fiber digestibility (Guedes *et al.*, 2008), and lower facal shedding of the zoonotic pathogen *E. coli* O157:H7 (Schamberger *et al.*, 2004).

In spite of their ability to elicit several beneficial effects, the performance of probiotics is often inconsistent and contradictory and depends on several factors such as the choice of microbial strain, diet, dosage level, duration of use, environment, and husbandry conditions (FAO, 2011c; Gaggìa *et al.*, 2010). The situation is further exacerbated by the fact that commercial products for animals often do not meet expected standards since the composition and viability of the strains may differ from those listed on the label (Wannaprasat *et al.*, 2009).

Nevertheless, the legislative frameworks in Canada, the EU, and the United States of America for the safety evaluation of probiotics have made significant progress (Gaggìa *et al.*, 2010). In the EU, subsequent to the ban on antimicrobials as growth promoters in 2006, the use of probiotics further expanded to maintain animal productivity (Vila *et al.*, 2010). Microorganisms utilized in animal feed in the EU are mainly bacterial strains of *Bacillus*, *Enterococcus*, *Lactobacillus*, *Pediococcus*, *Streptococcus*, and yeast strains belonging to the *S. cerevisiae* species and kluyveromyces (Anadón *et al.*, 2006). In addition to the aforementioned strains, currently approved microorganisms for animal feed application in the United States of America include strains of *Aspergillus*, *Bacteroides*, *Bifidobacterium*, and *Propionibacterium* (Flint and Garner, 2009).

Probiotics are also prevalent in a number of developing countries. In China, 15 microorganisms have been approved as feed additives and more than 400 companies are reported to be producing these additives (FAO, 2011c). The use of probiotics in commercial aquaculture in China has led to increased outputs and reduced costs (Qi *et al.*, 2009), and in India, the total market value of probiotics in aquaculture is estimated to be USD 109 million (Panigrahi and Azad, 2007).

K. Mutagenesis

Mutagenesis has been extensively used for the improvement of microorganisms of agricultural and industrial importance. Induced mutagenesis has been carried out to create starter strains with enhanced fermentation activity for enzyme (Awan *et al.*, 2011; Patel and Goyal, 2010; Ray, 2011; Xu *et al.*, 2011) and metabolite production (Demain and Adrio, 2008), or with resistance to viruses (FAO, 2011e). Site-specific mutagenesis allows targeted substitution, insertion or deletion of a single or a few base pairs (Blomqvist *et al.*, 2010) and has been applied to optimize properties such as increased thermostability of the enzyme of interest (Singh *et al.*, 2011b).

L. Transgenesis

Genetic modification, though common in developed countries for microbial strain improvement, is only now beginning to be applied for this purpose in developing countries. A few GM microorganisms have been approved in the food and beverage industry, for example, GM baker's yeast and brewers' yeast in the United Kingdom as well as two GM yeast strains for wine production in North America (FAO, 2011e). Many food processing enzymes in the United States of America are derived from GM microorganisms, for example, chymosin/rennin, phytase, and α-amylase (Olempska-Beer *et al.*, 2006). In developing countries, GM microorganisms are being used to produce enzymes in Argentina, Brazil, China, Cuba, and India; amino acids in China and Thailand; as well as nucleic acids and polysaccharides in China (FAO, 2011e). Recombinant bovine somatotropin, to increase milk yield, is produced by transgenic *E. coli* and is banned in Australia, Canada, the EU, Japan, and New Zealand but is approved in approximately 20 countries including the United States of America and some developing countries such as Brazil, Kenya, Mexico, and South Africa (FAO, 2011c).

VIII. CONCLUSIONS AND OUTLOOK

Developing countries are often endowed with a wealth of genetic resources. Agriculture is frequently an integral component of the economy and the harvest of wild populations, as in capture fisheries, forests, and bushmeat, is economically and culturally important. Yet they have not been able to harness this diversity of genetic resources to their full potential for many reasons, among them the lack of appropriate policies, limited human and institutional capacities, low R&D capacity and investment, inadequate infrastructure, and low levels of financial investments. The challenge, therefore, remains to manage GRFA effectively so as to conserve and enhance genetic diversity and at the same time sustainably utilize it for increased agricultural productivity as well as to ensure food security for the future.

Agricultural biotechnologies do not represent a "silver bullet," but they are an additional weapon in the arsenal to conserve GRFA and overcome constraints to agricultural production. Developments in biotechnology have been quite substantial in the past two decades and, as reviewed here, biotechnologies have made notable contributions and hold great promise for the management of GRFA. Molecular markers can be used in a variety of ways to characterize genetic resources, to identify priority genetic resources for conservation as well as to effectively manage *ex situ* collections; *in vitro* technologies offer complementary techniques to conventional conservation methods; technologies like tissue culture provide the means to overcome reproductive barriers; diagnostics and vaccines can assist in reducing economic losses due to debilitating diseases; and advances in "omic" technologies are playing an increasingly significant role in understanding the fundamental elements of plant and animal biology. Furthermore, biotechnologies can help to address the impact of emerging issues such as climate change and new diseases, on agriculture (FAO, 2011a–e).

While some biotechnologies such as AI and micropropagation have been widely adopted and applied in developing countries, uptake of other biotechnologies has been slower. Usually, the successful application of a given biotechnology depends on the presence of complementary factors (e.g., training and extension services), rather than on the effectiveness of the biotechnology *per se*. Additionally, non-GMO biotechnologies are often overshadowed by the debate on GMOs and there is a paucity of information/accurate assessments related to the application and potential socioeconomic effects of non-GMO biotechnologies (Sonnino *et al.*, 2009).

There is no one-size-fits-all solution, since there are substantial differences between sectors, species, regions, and countries. Moreover, within developing countries, there is considerable variation in funding and agricultural research capacities. Hence, decisions about what biotechnologies are appropriate and their subsequent development and adoption should be made carefully, based on reliable *ex ante* (e.g., sector-specific requirements and relevance to smallholders' needs) and *ex post* (e.g., adoption rate and genetic impact assessments) analyses, as well as suitability within existing development strategies.

Likewise, an enabling environment, with sound policies in place, is necessary to facilitate the appropriate application of biotechnologies for the management of GRFA. This includes appropriate IPR management, increased investments in the public sector, facilitation of public–private sector partnerships, improved market access, enhanced access to credit, tax subsidies and levies, creation of "biotechnology parks" (to colocate resources and capabilities), results-based management of research programs, and sharing technologies through collaboration platforms and initiatives, among others (FAO, 2011f–h).

The international technical conference on "Agricultural biotechnologies in developing countries: Options and opportunities in crops, forestry, livestock, fisheries and agro-industry to face the challenges of food insecurity and climate change" (ABDC-10),[189] convened by FAO in March 2010, brought together about 300 policy-makers, scientists, and representatives of intergovernmental and international nongovernmental organizations, including delegations from 42 FAO Member States. In the conference's keynote address, M.S.

[189]http://www.fao.org/biotech/abdc/.

Swaminathan succinctly stated "The bottom line of our national agricultural biotechnology policy should be the economic well being of farm families, food security of the nation, health security of the consumer, biosecurity of agriculture and health, protection of the environment and the security of national and international trade in farm commodities".

Indeed, at ABDC-10 the Member States reached a number of key conclusions, acknowledging, *inter alia*, that "Agricultural biotechnologies encompass a wide-range of tools and methodologies that are being applied to an increasing extent in crops, livestock, forestry, fisheries and aquaculture, and agro-industries, to help alleviate hunger and poverty, assist in adaptation to climate change and maintain the natural resource base, in both developing and developed countries" and agreeing, *inter alia*, that:

> "(a) Developing countries should significantly increase sustained investments in capacity building and development and safe use of biotechnologies; integrated with other agricultural technologies, including traditional knowledge, and maintain the natural resource base to support in particular, smallholders, producers and small biotechnology based enterprises; employing effective participatory approaches for the robust input from stakeholders in decision-making processes.
>
> (b) FAO and other relevant international organizations and donors should significantly increase their efforts to support the strengthening of national capacities in the development and appropriate use of pro-poor agricultural biotechnologies, and that they be directed to the needs of smallholders, consumers, producers and small biotechnology based enterprises in developing countries."

Acknowledgments

Comments from the following FAO colleagues, on parts or all of the document, are gratefully acknowledged: Devin Bartley, Paul Boettcher, Ruth Garcia Gomez, Kakoli Ghosh, Matthias Halwart, Kathrin Hett, Harinder Makkar, Dafydd Pilling, Rosa Rolle, John Ruane, Oudara Souvannavong, and Annika Wennberg.

We are especially thankful to the following external referees for providing their valuable inputs: Paolo Donini (International Potato Center, Peru), Ehsan Dulloo (Bioversity International, Italy), and E. M. Muralidharan (Kerala Forest Research Institute, India).

References

Abbott, C. L., Gilmore, S. R., Lewis, C. T., Chapados, J. T., Peters, R. D., Platt, H. W., Coffey, M. D., and Lévesque, C. A. (2010). Development of a SNP genetic marker system based on variation in microsatellite flanking regions of *Phytophthora infestans*. *Can. J. Plant Pathol.* **32,** 440–457.

Abril, N., Gion, J. M., Kerner, R., Müller-Starck, G., Cerrillo, R. M., Plomion, C., Renaut, J., Valledor, L., and Jorrin-Novo, J. V. (2011). Proteomics research on forest trees, the most recalcitrant and orphan plant species. *Phytochemistry* **72,** 1219–1242.

Adams, A., and Thompson, K. D. (2008). Recent applications of biotechnology to novel diagnostics for aquatic animals. *Rev. Sci. Tech.* **27,** 197–209.

Adams, A., and Thompson, K. D. (2011). Development of diagnostics for aquaculture: Challenges and opportunities. *Aquacult. Res.* **42,** 93–102.

Adams, A., Aoki, T., Berthe, F. C. J., Grisez, L., and Karunasagar, I. (2008). Recent technological advancements on aquatic animal health and their contributions toward reducing disease risks—A review. *In* "Diseases in Asian Aquaculture VI" (M. G. Bondad-Reantaso, C. V. Mohan, M. Crumlish, and R. P. Subasinghe, eds.), pp. 71–88. Fish Health Section, Asian Fisheries Society, Manila, Philippines.

Affolter, M., Grass, L., Vanrobaeys, F., Casado, B., and Kussmann, M. (2010). Qualitative and quantitative profiling of the bovine milk fat globule membrane proteome. *J. Proteomics* **73,** 1079–1088.

Agboh-Noameshie, A. R., Kinkingninhoun-Medagbe, F. M., and Diagne, A. (2007). Gendered impact of NERICA adoption on farmers' production and income in central Benin. *Paper presented at the International conference of African Association of Agricultural Economists (AAAE), 20-22 August 2007, Accra, Ghana.* .

Agrawal, G. K., and Rakwal, R. (2006). Rice proteomics: A cornerstone for cereal food crop proteomes. *Mass Spectrom. Rev.* **25,** 1–53.

Aguilar, C. N., Gutiérrez-Sánchez, G., Rado-Barragán, P. A., Rodríguez-Herrera, R., Martínez-Hernandez, J. L., and Contreras-Esquivel, J. C. (2008). Perspectives of solid state fermentation for production of food enzymes. *Am. J. Biochem. Biotechnol.* **4,** 354–366.

Agulleiro, M. J., Anguis, V., Cañavate, J. P., Martínez-Rodríguez, G., Mylonas, C. C., and Cerdá, J. (2006). Induction of spawning of captive-reared Senegal sole (*Solea senegalensis*) using different administration methods for gonadotropin-releasing hormone agonist. *Aquaculture* **257,** 511–524.

Ajmone-Marsan, P., Garcia, J. F., and Lenstra, J. A. (2010). On the origin of cattle: How aurochs became domestic and colonized the world. *Evol. Anthropol.* **19,** 148–157.

Akbari, M., Wenzl, P., Vanessa, C., Carling, J., Xia, L., Yang, S., Uszynski, G., Mohler, V., Lehmensiek, A., Kuchel, H., *et al.* (2006). Diversity Arrays Technology (DArT) for high-throughput profiling of the hexaploid wheat genome. *Theor. Appl. Genet.* **113,** 1409–1420.

Akkale, C., Yildirim, Z., Yildirim, M. B., Kaya, C., Ozturk, G., and Tanyolac, B. (2010). Assessing genetic diversity of some potato (*Solanum tuberosum* L.) genotypes grown in Turkey using the AFLP marker technique. *Turkish Journal of Field Crops* **15,** 73–78.

Alarcon, B., Vicedo, B., and Aznar, R. (2006). PCR-based procedures for detection and quantification of *Staphylococcus aureus* and their application in food. *J. Appl. Microbiol.* **100,** 352–364.

Alarcon, M. A., and Galina, C. S. (2009). Is embryo transfer a useful technique for small community farmers? FAO/IAEA International Symposium on Sustainable Improvement of Animal Production and Health, 8–10 June 2009, Synopses. pp. 152–153. IAEA Press, Vienna.

Alavi, S. M. H., and Cosson, J. (2006). Sperm motility in fishes. (II) Effects of ions and osmolality: A review. *Cell Biol. Int.* **30,** 1–14.

Ali, M. L., Rajewski, J. F., Baenziger, P. S., Gill, K. S., Eskridge, K. M., and Dweikat, I. (2008). Assessment of genetic diversity and relationship among a collection of US sweet sorghum germplasm by SSR markers. *Mol. Breed.* **21,** 497–509.

Alisi, C., Musella, R., Tasso, F., Ubaldi, C., Manzo, S., Cremisini, C., and Sprocati, A. R. (2009). Bioremediation of diesel oil in a cocontaminated soil by bioaugmentation with a microbial formula tailored with native strains selected for heavy metals resistance. *Sci. Total Environ.* **407,** 3024–3032.

Allendorf, F. W., Hohenlohe, P. A., and Luikart, G. (2010). Genomics and the future of conservation genetics. *Nat. Rev. Genet.* **11,** 697–709.

Almeida, A. M., Campos, A., Francisco, R., van Harten, S., Cardoso, L. A., and Coelho, A. V. (2010). Proteomic investigation of the effects of weight loss in the gastrocnemius muscle of wild and NZW rabbits via 2D-electrophoresis and MALDI-TOF MS. *Anim. Genet.* **41,** 260–272.

Alonso, P., Moncaleán, P., Fernández, B., Rodríguez, A., Centeno, M. L., and Ordás, R. J. (2006). An improved micropropagation protocol for stone pine (*Pinus pinea* L.). *Ann. For. Sci.* **63,** 879–885.

Alpuerto, V. L. E. B., Norton, G. W., Alwang, J., and Ismail, A. M. (2009). Economic impact analysis of marker-assisted breeding for tolerance to salinity and phosphorous deficiency in rice. *Appl. Econ. Perspect. Policy* **31,** 779–792.

Altinok, I., and Kurt, I. (2003). Molecular diagnosis of fish diseases: A review. *Turk. J. Fish. Aquat. Sci.* **3,** 131–138.

Aluru, N., and Vijayan, M. M. (2009). Stress transcriptomics in fish: A role for genomic cortisol signaling. *Gen. Comp. Endocrinol.* **164,** 142–150.

Álvarez, I., Royo, L. J., Gutiérrez, J. P., Fernández, I., Arranz, J. J., and Goyache, F. (2008). Relationship between genealogical and microsatellite information characterising losses of genetic variability: Empirical evidence from the rare Xalda sheep breed. *Livest. Sci.* **115,** 80–88.

Álvarez, J. M., Majada, J., and Ordás, R. J. (2009). An improved micropropagation protocol for maritime pine (*Pinus pinaster* Ait.) isolated cotyledons. *Forestry* **82,** 175–184.

Amigues, Y., Boitard, S., Bertrand, C., SanCristobal, M., and Rocha, D. (2011). Genetic characterization of the Blonde d'Aquitaine cattle breed using microsatellite markers and relationship with three other French cattle populations. *J. Anim. Breed. Genet.* **128,** 201–208.

Ammar, K., Mergoum, M., and Rajaram, S. (2004). The history and evolution of Triticale. *In* "Triticale Improvement and Production" (M. Mergoum and H. Gomez-Macpherson, eds.), pp. 1–9. FAO, Rome.

Anadón, A., Martinez-Larrañaga, M. R., and Martinez, M. A. (2006). Probiotics for animal nutrition in the European Union. Regulation and safety assessment. *Regul. Toxicol. Pharmacol.* **45,** 91–95.

Andersen, S. B. (2005). Haploids in the improvement of woody species. *In* "Haploids in Crop Improvement II" (C. E. Palmer, W. A. Keller, and K. J. Kasha, eds.), Haploids in Crop Improvement II, Vol. 56, pp. 243–257. Springer, London.

Andersson, M. S., and de Vicente, M. C. (2010). Gene Flow Between Crops and Their Wild Relatives. Johns Hopkins University Press, Baltimore.

Angel, F., Barney, V. E., Tohme, J., and Roca, W. M. (1996). Stability of cassava plants at the DNA level after retrieval from 10 years of *in vitro* storage. *Euphytica* **90,** 307–313.

Angienda, P. O., Lee, H. J., Elmer, K. R., Abila, R., Waindi, E. N., and Meyer, A. (2011). Genetic structure and gene flow in an endangered native tilapia fish (*Oreochromis esculentus*) compared to invasive Nile tilapia (*Oreochromis niloticus*) in Yala swamp, East Africa. *Conserv. Genet.* **12,** 243–255.

Araki, H., and Schmid, C. (2010). Is hatchery stocking a help or harm? Evidence, limitations and future directions in ecological and genetic surveys. *Aquaculture* **308,** S2–S11.

Araújo, A. M., Guimarães, S. E. F., Pereira, C. S., Lopes, P. S., Rodrigues, M. T., and Machado, T. M. M. (2010). Paternity in Brazilian goats through the use of DNA microsatellites. *R. Bras. Zootec.* **39,** 1011–1014.

Aravanopoulos, F. A. (2010). Clonal identification based on quantitative, codominant, and dominant marker data: A comparative analysis of selected willow (*Salix* L.) clones. *Int. J. For. Res.* **2010,** 1–8. doi:10.1155/2010/906310.

Arias, J. A., Keehan, M., Fisher, P., Coppieters, W., and Spelman, R. (2009). A high density linkage map of the bovine genome. *BMC Genet.* **10,** 18.

Arya, I. D., Sharma, S., Chauhan, S., and Arya, S. (2009). Micropropagation of superior eucalyptus hybrids FRI-5 (*Eucalyptus camaldulensis* Dehn x *E. tereticornis* Sm) and FRI-14 (*Eucalyptus torelliana* F.V. Muell x *E. citriodora* Hook): A commercial multiplication and field evaluation. *Afr. J. Biotechnol.* **8,** 5718–5726.

Atmakuri, A. R., Chaudhury, R., Malik, S. K., Kumar, S., Ramachandran, R., and Qadri, S. M. H. (2009). Mulberry biodiversity conservation through cryopreservation. *In Vitro Cell. Dev. Biol. Plant* **45,** 639–649.

Attwood, G. T., Altermann, E., Kelly, W. J., Leahy, S. C., Zhang, L., and Morrison, M. (2011). Exploring rumen methanogen genomes to identify targets for methane mitigation strategies. *Anim. Feed Sci. Technol.* **166–167,** 65–75.

Aubin-Horth, N., Landry, C. R., Letcher, B. H., and Hofmann, H. A. (2005). Alternative life histories shape brain gene expression profiles in males of the same population. *Proc. Biol. Sci.* **272,** 1655–1662.

Avery, T. S., Boyce, D., and Brown, J. A. (2004). Mortality of yellowtail flounder, *Limanda ferruginea* (Storer), eggs: Effects of temperature and hormone-induced ovulation. *Aquaculture* **230,** 297–311.

Awan, M. S., Tabbasam, N., Ayub, N., Babar, M. E., Mehboob-ur-Rahman, Rana, S. M., and Rajoka, M. I. (2011). Gamma radiation induced mutagenesis in *Aspergillus niger* to enhance its microbial fermentation activity for industrial enzyme production. *Mol. Biol. Rep.* **38,** 1367–1374.

Azaiez, A., Boyle, B., Levée, V., and Séguin, A. (2009). Transcriptome profiling in hybrid poplar following interactions with Melampsora rust fungi. *Mol. Plant Microbe Interact.* **22,** 190–200.

Babiak, I., Dobosz, S., Goryczko, K., Kuzminski, H., Brzuzan, P., and Ciesielski, S. (2002). Androgenesis in rainbow trout using cryopreserved spermatozoa: The effect of processing and biological factors. *Theriogenology* **57,** 1229–1249.

Babu, R., Nair, S. K., Kumar, A., Venkatesh, S., Sekhar, J. C., Singh, N. N., Srinivasan, G., and Gupta, H. S. (2005). Two-generation marker-aided backcrossing for rapid conversion of normal maize lines to quality protein maize (QPM). *Theor. Appl. Genet.* **111,** 888–897.

Bailey, K. L., Boyetchko, S. M., and Längle, T. (2010). Social and economic drivers shaping the future of biological control: A Canadian perspective on the factors affecting the development and use of microbial biopesticides. *Biol. Control* **52,** 221–229.

Baker, C. S. (2008). A truer measure of the market: The molecular ecology of fisheries and wildlife trade. *Mol. Ecol.* **17,** 3985–3998.

Baker, M. E., Ruggeri, B., Sprague, L. J., Eckhardt-Ludka, C., Lapira, J., Wick, I., Soverchia, L., Ubaldi, M., Polzonetti-Magni, A. M., Vidal-Dorsch, F., *et al.* (2009). Analysis of endocrine disruption in Southern California coastal fish using an aquatic multispecies microarray. *Environ. Health Perspect.* **117,** 223–230.

Balamurugan, V., Venkatesan, G., Sen, A., Annamalai, L., Bhanuprakash, V., and Singh, R. K. (2010). Recombinant protein-based viral disease diagnostics in veterinary medicine. *Expert Rev. Mol. Diagn.* **10,** 731–753.

Balcazar, J. L., de Blas, I., Ruiz-Zarzuela, I., Cunningham, D., Vendrell, D., and Múzquiz, J. L. (2006). The role of probiotics in aquaculture. *Vet. Microbiol.* **114,** 173–186.

Balcazar, J. L., Vendrell, D. L., deBlas, I., Ruiz-Zarzuela, I., and Muzquiz, J. L. (2009). Effect of *Lactococcus lactis* CLFP 100 and *Leuconostoc mesenteroides* CLFP 196 on *Aeromonas salmonicida* infection in brown trout (*Salmo trutta*). *J. Mol. Microbiol. Biotechnol.* **17,** 153–157.

Balfourier, F., Roussel, V., Strelchenko, P., Exbrayat-Vinson, F., Sourdille, P., Boutet, G., Koenig, J., Ravel, C., Mitrofanova, O., Beckert, M., *et al.* (2007). A worldwide bread wheat core collection arrayed in a 384-well plate. *Theor. Appl. Genet.* **114**(7), 1265–1275.

Banat, I. M., Franzetti, A., Gandolfi, I., Bestetti, G., Martinotti, M. G., Fracchia, L., Smyth, T. J., and Marchant, R. (2010). Microbial biosurfactants production, applications and future potential. *Appl. Microbiol. Biotechnol.* **87,** 427–444.

Bannantine, J. P., and Talaat, A. M. (2010). Genomic and transcriptomic studies in *Mycobacterium avium* subspecies *paratuberculosis*. *Vet. Immunol. Immunopathol.* **138**(4), 303–311.

Barakat, A., DiLoreto, D. S., Zhang, Y., Smith, C., Baier, K., Powell, W. A., Wheeler, N., Sederoff, R., and Carlson, J. E. (2009). Comparison of the transcriptomes of American chestnut (*Castanea dentata*) and Chinese chestnut (*Castanea mollissima*) in response to the chestnut blight infection. *BMC Plant Biol.* **9,** 51.

Baranski, M., Moen, T., and Våge, D. I. (2010). Mapping of quantitative trait loci for flesh colour and growth traits in Atlantic salmon (*Salmo salar*). *Genet. Sel. Evol.* **42,** 17.

Barbazuk, W. B., Emrich, S. J., Chen, H. D., Li, L., and Schnable, P. S. (2007). SNP discovery via 454 transcriptome sequencing. *Plant J.* **51,** 910–918.

Barkley, N. A., and Wang, M. L. (2008). Application of TILLING and EcoTILLING as reverse genetic approaches to elucidate the function of genes in plants and animals. *Curr. Genomics* **9,** 212–226.

Barnard, R. T. (2010). Recombinant vaccines. *Expert Rev. Vaccines* **9,** 461–463.

Bartley, D., Bagley, M., Gall, G., and Bentley, B. (1992). Use of linkage disequilibrium data to estimate effective size of hatchery and natural fish populations. *Conserv. Biol.* **6,** 365–375.

Bartley, D. M., Benzie, J. A. H., Brummett, R. E., Davy, F. B., De Silva, S. S., Eknath, A. E., Guo, X., Halwart, M., Harvey, B., Jeney, Z., *et al.* (2009). The use and exchange of aquatic genetic resources for food and agriculture. *Background Study Paper No. 45.* ftp://ftp.fao.org/docrep/fao/meeting/017/ak527e.pdf.

Bautista, G. H., Cruz, H. A., Nesme, X., Valdés, M., Mendoza, H. A., and Fernandez, M. P. (2011). Genomospecies identification and phylogenomic relevance of AFLP analysis of isolated and non-isolated strains of *Frankia* spp. *Syst. Appl. Microbiol.* **34,** 200–206.

Beacham, T. D., Wetklo, M., Deng, L., and MacConnachie, C. (2011). Coho salmon population structure in North America determined from microsatellites. *Trans. Am. Fish. Soc.* **140,** 253–270.

Beaumont, A., Boudry, P., and Hoare, K. (2010). Biotechnology and Genetics in Fisheries and Aquaculture. Wiley-Blackwell, Oxford.

Beer, M., Reimann, I., Hoffmann, B., and Depner, K. (2007). Novel marker vaccines against classical swine fever. *Vaccine* **25,** 5665–5670.

Behl, R., Sheoran, N., Behl, J., and Vijh, R. K. (2006). Genetic analysis of Ankamali pigs of India using microsatellite markers and their comparison with other domesticated Indian pig types. *J. Anim. Breed. Genet.* **123,** 131–135.

Beier, S., Witzel, K.-P., and Marxsen, J. (2008). Bacterial community composition in central European running waters examined by TGGE and sequence analysis of 16S rDNA. *Appl. Environ. Microbiol.* **74,** 188–199.

Bekkevold, D., Hansen, M. M., and Nielsen, E. E. (2006). Genetic impact of gadoid culture on wild fish populations: Predictions, lessons from salmonids, and possibilities for minimizing adverse effects. *ICES J. Mar. Sci.* **63,** 198–208.

Belyaeva, M., and Taylor, D. J. (2009). Cryptic species within the *Chydorus sphaericus* species complex (Crustacea: Cladocera) revealed by molecular markers and sexual stage morphology. *Mol. Phylogenet. Evol.* **50,** 534–546.

Ben-David, E. A., Zaady, E., Sher, Y., and Nejidat, A. (2011). Assessment of the spatial distribution of soil microbial communities in patchy arid and semi-arid landscapes of the Negev Desert using combined PLFA and DGGE analyses. *FEMS Microbiol. Ecol.* **76,** 492–503.

Bendixen, E., Danielsen, M., Hollung, K., Gianazza, E., and Miller, I. (2011). Farm animal proteomics—A review. *J. Proteomics* **74,** 282–293.

Bennewitz, J., Reinsch, N., Reinhardt, F., Liu, Z., and Kalm, E. (2004). Top down preselection using marker assisted estimated of breeding values in dairy cattle. *J. Anim. Breed. Genet.* **121,** 307–318.

Berger, R. G. (2009). Biotechnology of flavours—The next generation. *Biotechnol. Lett.* **31,** 1651–1659.

Bernardo, R. (2008). Molecular markers and selection for complex traits in plants: Learning from the past 20 years. *Crop Sci.* **48,** 1649–1664.

Berntson, E. A., and Moran, P. (2009). The utility and limitations of genetic data for stock identification and management of North Pacific rockfish (*Sebastes* spp.). *Rev. Fish Biol. Fish.* **19,** 233–247.

Berthouly, C., Leroy, G., Nhu Van, T., Hoang Thanh, H., Bed'Hom, B., Nguyen, B. T., Vu Chi, C., Monicat, F., Tixier-Boichard, M., Verrier, E., *et al.* (2009). Genetic analysis of local Vietnamese chickens provides evidence of gene flow from wild to domestic populations. *BMC Genet.* **10,** 1.

Besnard, G., Baradat, P., Breton, C., Khadari, B., and Berville, A. (2001). Olive domestication from structure of oleasters and cultivars using RAPDs and mitochondrial RFLP. *Genet. Sel. Evol.* **33** (Suppl. 1), S251–S268.

Bhattacharjee, R. B., Singh, A., and Mukhopadyay, S. N. (2008). Use of nitrogen-fixing bacteria as biofertilizer for non-legumes: Prospects and challenges. *Appl. Microbiol. Biotechnol.* **80,** 199–209.

Bickford, D., Lohman, D. J., Sodhi, N. S., Ng, P. K. L., Meier, R., Winker, K., Ingram, K. K., and Das, I. (2007). Cryptic species as a window on diversity and conservation. *Trends Ecol. Evol.* **22,** 148–155.

Bioversity International (2007). Guidelines for the development of crop descriptor lists. Bioversity Technical Bulletin Series. Bioversity International, Rome. http://cropgenebank.sgrp.cgiar.org/images/file/learning_space/technicalbulletin13.pdf.

Bjarnadottir, S. G., Hollung, K., Faergestad, E. M., and Veiseth-Kent, E. (2010). Proteome changes in bovine longissimus thoracis muscle during the first 48 h postmortem: Shifts in energy status and myofibrillar stability. *J. Agric. Food Chem.* **58,** 7408–7414.

Blackie, C. T., Morrissey, M. B., Danzmann, R. G., and Ferguson, M. M. (2011). Genetic divergence among broodstocks of Arctic charr *Salvelinus alpinus* in eastern Canada derived from the same founding populations. *Aquacult. Res.* **42,** 1440–1452.

Blair, M. W., González, L. F., Kimani, P. M., and Butare, L. (2010). Genetic diversity, inter-gene pool introgression and nutritional quality of common beans (*Phaseolus vulgaris* L.) from Central Africa. *Theor. Appl. Genet.* **121,** 237–248.

Blanchet, S., Páez, D. J., Bernatchez, L., and Dodson, J. J. (2008). An integrated comparison of captive-bred and wild Atlantic salmon (*Salmo salar*): Implications for supportive breeding programs. *Biol. Conserv.* **141,** 1989–1999.

Blomqvist, T., Steinmoen, H., and Havarstein, L. S. (2010). A food-grade site-directed mutagenesis system for *Streptococcus thermophilus* LMG 18311. *Lett. Appl. Microbiol.* **50,** 314–319.

Boa, E. (2010). Plant healthcare for poor farmers around the world: Gathering demand and innovative responses. *In* "Knowledge and Technology Transfer for Plant Pathology" (N. V. Hardwick and M. L. Gullino, eds.), Plant Pathology in the 21st Century, Vol. 4, pp. 1–16. Springer, Dordrecht, Heidelberg, London, New York.

Boa-Amponsem, K., and Minozzi, G. (2006). The state of development of biotechnologies as they relate to the management of animal genetic resources and their potential application in developing countries. *Background Study Paper 33.* ftp://ftp.fao.org/docrep/fao/meeting/015/j8959e.pdf.

Bodzsar, N., Eding, H., Revay, T., Hidas, A., and Weigend, S. (2009). Genetic diversity of Hungarian indigenous chicken breeds based on microsatellite markers. *Anim. Genet.* **40,** 1–7.

Boettcher, P. J., Stella, A., Pizzi, F., and Gandini, G. (2005). The combined use of embryos and semen for cryogenic conservation of mammalian livestock genetic resources. *Genet. Sel. Evol.* **37,** 657–675.

Boettcher, P. J., Tixier-Boichard, M., Toro, M. A., Simianer, H., Eding, H., Gandini, G., Joost, S., Garcia, D., Colli, L., Ajmone-Marsan, P., *et al.* (2010). Objectives, criteria and methods for using molecular genetic data in priority setting for conservation of animal genetic resources. *Anim. Genet.* **41,** 64–77.

Boichard, D., Fritz, S., Rossignol, M. N., Boscher, M. Y., Malafosse, A., and Colleau, J. J. (2002). Implementation of marker-assisted selection in French dairy cattle. *Proc. 7th World Cong. Genet. Appl. Livest. Prod.* **33,** 19–22.

Bonants, P., Groenewald, E., Rasplus, J. Y., Maes, M., de Vos, P., Frey, J., Boonham, N., Nicolaisen, M., Bertacini, A., Robert, V., *et al.* (2010). QBOL: A new EU project focusing on DNA barcoding of Quarantine organisms. *Bull. OEPP* **40,** 30–33.

Bondad-Reantaso, M. G., Subasinghe, R. P., Arthur, J. R., Ogawa, K., Chinabut, S., Adlard, R., Tan, Z., and Shariff, M. (2005). Disease and health management in Asian aquaculture. *Vet. Parasitol.* **132,** 249–272.

Bonfante, P., and Genre, A. (2010). Mechanisms underlying beneficial plant—Fungus interactions in mycorrhizal symbiosis. *Nat. Commun.* **1,** 48.

Bonga, J. M., Klimaszewska, K. K., and von Aderkas, P. (2010). Recalcitrance in clonal propagation, in particular of conifers. *Plant Cell Tissue Organ Cult.* **100,** 241–254.

Boonkerd, N. (2002). Development of inoculant production and utilisation in Thailand. *In* "Inoculants and Nitrogen Fixation of Legumes in Vietnam" (D. Herridge, ed.), Proceedings Workshop held in Hanoi, Vietnam, February 17–18, 2001. pp. 95–104.

Boonthai, T., Vuthiphandchai, V., and Nimrat, S. (2011). Probiotic bacteria effects on growth and bacterial composition of black tiger shrimp (*Penaeus monodon*). *Aquacult. Nutr.* **17,** 634–644.

Borneman, A. R., Bartowsky, E. J., McCarthy, J., and Chambers, P. J. (2010). Genotypic diversity in *Oenococcus oeni* by high-density microarray comparative genome hybridization and whole genome sequencing. *Appl. Microbiol. Biotechnol.* **86,** 681–691.

Börner, A., Chebotar, S., and Korzun, V. (2000). Molecular characterization of the genetic integrity of wheat (*Triticum aestivum* L.) germplasm after long-term maintenance. *Theor. Appl. Genet.* **100,** 494–497.

Boswell, M. G., Wells, M. C., Kirk, L. M., Ju, Z., Zhang, Z., Booth, R. E., and Walter, R. B. (2009). Comparison of gene expression responses to hypoxia in viviparous (*Xiphophorus*) and oviparous (*Oryzias*) fishes using a medaka microarray. *Comp. Biochem. Physiol. C Toxicol. Pharmacol.* **149,** 258–265.

Bott, K., Kornely, G. W., Donofrio, M. C., Elliott, R. F., and Scribner, K. T. (2009). Mixed-stock analysis of lake sturgeon in the Menominee River sport harvest and adjoining waters of Lake Michigan. *N. Am. J. Fish. Manag.* **29,** 1636–1643.

Bozinovic, G., and Oleksiak, M. F. (2011). Genomic approaches with natural fish populations from polluted environments. *Environ. Toxicol. Chem.* **30,** 283–289.

Braley, R. D. (1985). Serotonin-induced spawning in giant clams (Bivlavia: Tridacnidae). *Aquaculture* **47,** 321–325.

Brar, D. (2005). Broadening the genepool and exploiting heterosis in cultivated rice. *In* "Rice is Life: Scientific Perspectives for the 21st Century" (K. Toriyama, K. L. Heong, and B. Hardy, eds.)., Proceedings of the World Rice Research Conference, Tokyo and Tsukuba, Japan, 4–7 November 2004.

Breton, C., Tersac, M., and Bervillé, A. (2006). Genetic diversity and gene flow between the wild olive (oleaster, *Olea europaea* L.) and the olive: Several Plio-Pleistocene refuge zones in the Mediterranean basin suggested by simple sequence repeats analysis. *J. Biogeogr.* **33,** 1916–1928.

Breton, C., Terral, J. F., Pinatel, C., Médail, F., Bonhomme, F., and Bervillé, A. (2009). The origins of the domestication of the olive tree. *C. R. Biol.* **332,** 1059–1064.

Broders, K. D., Woeste, K. E., Sanmiguel, P. J., Westerman, R. P., and Boland, G. J. (2011). Discovery of single-nucleotide polymorphisms (SNPs) in the uncharacterized genome of the ascomycete *Ophiognomonia clavigignenti-juglandacearum* from 454 sequence data. *Mol. Ecol. Resour.* **11,** 693–702.

Broeck, A. V., Stormeb, V., Cottrellc, J. E., Boerjanb, W., Van Bockstaeled, E., Quataerta, P., and Van Slycken, J. (2004). Gene flow between cultivated poplars and native black poplar (*Populus nigra* L.): A case study along the river Meuse on the Dutch–Belgian border. *For. Ecol. Manage.* **197,** 307–310.

Broeck, A. V., Villar, M., Bockstaele, E. V., and VanSlycken, J. (2005). Natural hybridization between cultivated poplars and their wild relatives: Evidence and consequences for native poplar populations. *Ann. For. Sci.* **62,** 601–613.

Brown, G. B., Bassoni, D. L., Gil, G. P., Fontana, J. R., Wheeler, N. C., Megraw, R. A., Davis, M. F., Sewell, M. M., Tuskan, G. A., and Neale, D. B. (2003). Identification of quantitative trait loci influencing wood property traits in loblolly pine (*Pinus taeda* L.). III. QTL verification and candidate gene mapping. *Genetics* **164,** 1537–1546.

Brown, R. C., Woolliams, J. A., and McAndrew, B. J. (2005). Factors influencing effective population size in commercial populations of gilthead seabream, *Sparus aurata. Aquaculture* **247,** 219–225.

Budak, H., Bölek, Y., Dokuyucu, T., and Akkaya, A. (2004). Potential uses of molecular markers in crop improvement. *KSU J. Sci. Eng.* **7,** 75–79.

Buddle, B. M., Denis, M., Attwood, G. T., Altermann, E., Janssen, P. H., Ronimus, R. S., Pinares-Patiño, C. S., Muetzel, S., and Wedlock, D. N. (2011). Strategies to reduce methane emissions from farmed ruminants grazing on pasture. *Vet. J.* **188,** 11–17.

Bundock, P. C., Eliott, F. G., Ablett, G., Benson, A. D., Casu, R. E., Aitken, K. S., and Henry, R. J. (2009). Targeted single nucleotide polymorphism (SNP) discovery in a highly polyploidy plant species using 454 sequencing. *Plant Biotechnol. J.* **7,** 347–354.

Burczyk, J., DiFazio, S. P., and Adams, W. T. (2004). Gene flow in forest trees: How far do genes really travel? *Forest Genet.* **11,** 179–192.

Burgain, J., Gaiani, C., Linder, M., and Scher, J. (2011). Encapsulation of probiotic living cells: From laboratory scale to industrial applications. *J. Food Eng.* **104,** 467–483.

Burgarella, C., Lorenzo, Z., Jabbour-Zahab, R., Lumaret, R., Guichoux, E., Petit, R. J., Soto, Á., and Gil, L. (2009). Detection of hybrids in nature: Application to oaks (*Quercus suber* and *Q. ilex*). *Heredity* **102,** 442–452.

Burger, J. C., Chapman, M. A., and Burke, J. M. (2008). Molecular insights into the evolution of crop plants. *Am. J. Bot.* **95,** 113–122.

Burridge, L., Weis, J. S., Cabello, F., Pizarro, J., and Bostick, K. (2010). Chemical use in salmon aquaculture: A review of current practices and possible environmental effects. *Aquaculture* **306,** 7–23.

Busov, V., Fladung, M., Groover, A., and Strauss, S. (2005). Insertional mutagenesis in *Populus*: Relevance and feasibility. *Tree Genet. Genomes* **1,** 135–142.

Busov, V., Yordanov, Y., Gou, J., Meilanm, R., Ma, C., Regan, S., and Strauss, S. (2011). Activation tagging is an effective gene tagging system in *Populus*. *Tree Genet. Genomes* **7,** 91–101.

Caballero, A. (1994). Developments in the prediction of effective population size. *Heredity* **73,** 657–679.

Cabrita, E., Sarasquete, C., Martínez-Páramo, S., Robles, V., Beirão, J., Pérez-Cerezales, S., and Herráez, M. P. (2010). Cryopreservation of fish sperm: Applications and perspectives. *J. Appl. Ichthyol.* **26,** 623–635.

Caffey, R. H., and Tiersch, T. R. (2000). Cost analysis for integrating cryopreservation into an existing fish hatchery. *J. World Aquac. Soc.* **31,** 51–58.

Caicedo, A. L., Williamson, S. H., Hernandez, R. D., Boyko, A., Fledel-Alon, A., York, T. L., Polato, N. R., Olsen, K. M., Nielsen, R., McCouch, S. R., *et al.* (2007). Genome-wide patterns of nucleotide polymorphism in domesticated rice. *PLoS Genet.* **3,** 1289–1299.

Cal, R. M., Vidal, S., Gómez, C., Álvarez-Blázquez, B., Martínez, P., and Piferrer, F. (2006). Growth and gonadal development in diploid and triploid turbot (*Scophthalmus maximus*). *Aquaculture* **251,** 99–108.

Caliz, J., Vila, X., Marti, E., Sierra, J., Cruanas, R., Garau, M. A., and Montserrat, G. (2011). Impact of chlorophenols on microbiota of an unpolluted acidic soil: Microbial resistance and biodegradation. *FEMS Microbiol. Ecol.* **78,** 150–164.

Cañón, J., García, D., García-Atance, M. A., Obexer-Ruff, G., Lenstra, J. A., Ajmone-Marsan, P., Dunner, S., and the ECONOGENE Consortium (2006). Geographical partitioning of goat diversity in Europe and the Middle East. *Anim. Genet.* **37,** 327–334.

Cardarelli, M., Rouphael, Y., Rea, E., and Colla, G. (2010). Mitigation of alkaline stress by arbuscular mycorrhiza in zucchini plants grown under mineral and organic fertilization. *J. Plant Nutr. Soil Sci.* **173,** 778–787.

Carrapiço, F., Teixeira, G., and Diniz, M. A. (2000). *Azolla* as a biofertiliser in Africa. A challenge for the future. *Rev. Ciências Agrárias* **23,** 120–138.

Carwell, D. B., Pitchford, J. A., Gentry, G. T., Blackburn, H. D., Bondioli, K. R., and Godke, R. A. (2009). 19 Beef cattle pregnancy rates following insemination with aged frozen angus semen. *J. Reprod. Fertil. Dev.* **22,** 167–168.

Casasoli, M., Derory, J., Morera-Dutrey, C., Brendel, O., Porth, I., Guehl, J.-M., Villani, F., and Kremer, A. (2006). Comparison of quantitative trait loci for adaptive traits between oak and chestnut based on an expressed sequence tag consensus map. *Genetics* **172,** 533–546.

Case, R. J., Boucher, Y., Dahllöf, I., Holmström, C., Doolittle, W. F., and Kjelleberg, S. (2007). Use of 16S rRNA and *rpoB* genes as molecular markers for microbial ecology studies. *Appl. Environ. Microbiol.* **73,** 278–288.

Castaño-Sánchez, C., Fuji, K., Ozaki, A., Hasegawa, O., Sakamoto, T., Morishima, K., Nakayama, I., Fujiwara, A., Masaoka, T., Okamoto, H., *et al.* (2010). A second generation genetic linkage map of Japanese flounder (*Paralichthys olivaceus*). *BMC Genomics* **11,** 554.

Castillo, A. G. F., Beall, E., Moran, P., Martinez, J. L., Ayllon, F., and García-Vázquez, E. (2007). Introgression in the genus *Salmo* via allotriploids. *Mol. Ecol.* **16,** 1741–1748.

Castro, J., Pino, A., Hermida, M., Bouza, C., Riaza, A., Ferreiro, I., Sanchez, L., and Martinez, P. (2006). A microsatellite marker tool for parentage analysis in senegal sole (*Solea senegalensis*): Genotyping errors, null alleles and conformance to theoretical assumptions. *Aquaculture* **261,** 1194–1203.

Caswell, K. L., and Kartha, K. K. (2009). Recovery of plants from pea and strawberry meristems cryopreserved for 28 years. *Cryo Letters* **30,** 41–46.

Cavileer, T., Hunter, S., Okutsu, T., Yoshizaki, G., and Nagler, J. J. (2009). Identification of novel genes associated with molecular sex differentiation in the embryonic gonads of rainbow trout (*Oncorhynchus mykiss*). *Sex. Dev.* **3,** 214–224.

CBOL Plant Working Group (2009). A DNA barcode for land plants. *Proc. Natl. Acad. Sci. U.S.A.* **106,** 12794–12797.

CBSG (2011). *Working Group Report: Review and Reconsideration of the IUCN Technical Guidelines on the Management of Ex Situ Populations for Conservation: Why, when (and how) to establish an ex situ population. Proceedings from the 2010 CBSG Annual Meeting, Cologne, Germany.* http://www.cbsg.org/cbsg/content/files/2010_Ann_Mtg_Newsletter/iucn_technical_guidelines.pdf.

Cervera, M. T., Storme, V., Soto, A., Ivens, B., Van Montagu, M., Rajora, O. P., and Boerjan, W. (2005). Intraspecific and interspecific genetic and phylogenetic relationships in the genus *Populus* based on AFLP markers. *Theor. Appl. Genet.* **111,** 1440–1456.

Chacón, M. I., Pickersgill, S. B., and Debouck, D. G. (2005). Domestication patterns in common bean (*Phaseolus vulgaris* L.) and the origin of the Mesoamerican and Andean cultivated races. *Theor. Appl. Genet.* **110,** 432–444.

Chagné, D., Brown, G. R., Lalanne, C., Madur, D., Pot, D., Neale, D., and Plomion, C. (2003). Comparative genome and QTL mapping between maritime and loblolly pines. *Mol. Breed.* **12,** 185–195.

Chandler, D., Davidson, G., Grant, W. P., Greaves, J., and Satchel, G. M. (2008). Microbial biopesticides for integrated crop management: An assessment of environmental and regulatory sustainability. *Trends Food Sci. Technol.* **19,** 275–283.

Chandrika, M., and Rai, V. R. (2009). Genetic fidelity in micropropagated plantlets of *Ochreinauclea missionis* an endemic, threatened and medicinal tree using ISSR markers. *Afr. J. Biotechnol.* **8,** 2933–2938.

Chapman, M. A., Pashley, C. H., Wenzler, J., Hvala, J., Tang, S., Knapp, S. J., and Burke, J. M. (2008). A genomic scan for selection reveals candidates for genes involved in the evolution of cultivated sunflower (*Helianthus annuus*). *Plant Cell* **20,** 2931–2945.

Chapman, D. D., Pinhal, D., and Shivji, M. S. (2009). Tracking the fin trade: Genetic stock identification in western Atlantic scalloped hammerhead sharks *Sphyrna lewini*. *Endanger. Species Res.* **9,** 221–228.

Chávez-Crooker, P., and Obreque-Contreras, J. (2010). Bioremediation of aquaculture wastes. *Curr. Opin. Biotechnol.* **21,** 313–317.

Chaze, T., Meunier, B., Chambon, C., Jurie, C., and Picard, B. (2008). *In vivo* proteome dynamics during early bovine myogenesis. *Proteomics* **8,** 4236–4248.

Chen, L. J., Lee, D. S., Song, Z. P., Suh, H. S., and Lu, B. (2004). Gene Flow from cultivated rice (*Oryza sativa*) to its weedy and wild relatives. *Ann. Bot.* **93,** 67–73.

Chen, F., Lee, Y., Jiang, Y., Wang, S., Peatman, E., Abernathy, J., Liu, H., Liu, S., Kucuktas, H., Ke, C., *et al.* (2010). Identification and characterization of full-length cDNAs in channel catfish (*Ictalurus punctatus*) and blue catfish (*Ictalurus furcatus*). *PLoS One* **5,** e11546.

Chen, C. Y., Misztal, I., Aguilar, I., Tsuruta, S., Meuwissen, T. H. E., Aggrey, S. E., Wing, T., and Muir, W. M. (2011). Genome-wide marker-assisted selection combining all pedigree phenotypic information with genotypic data in one step: An example using broiler chickens. *J. Anim. Sci.* **89,** 23–28.

Cheng, J. J., and Timilsina, G. R. (2010). Advanced biofuel technologies: Status and barriers. *Policy Research Working Paper 5411*, The World Bank. http://go.worldbank.org/663HDWL6Q0.

Chereguini, O., Banda, I., Rasines, I., and Fernandez, A. (2001). Larval growth of turbot, *Scophthalmus maximus* (L.) produced with fresh and cryopreserved sperm. *Aquacult. Res.* **32,** 133–143.

Chew, P. C., Abd-Rashid, Z., Hassan, R., Asmuni, M., and Chuah, H. P. (2010). Semen cryo-bank of the Malaysian Mahseer (*Tor tambroides* and *T. douronensis*). *J. Appl. Ichthyol.* **26,** 726–731.

Childers, C. P., Reese, J. T., Sundaram, J. P., Vile, D. C., Dickens, C. M., Childs, K. L., Salih, H., Bennett, A. K., Hagen, D. E., Adelson, D. L., *et al.* (2011). Bovine Genome Database: Integrated tools for genome annotation and discovery. *Nucleic Acids Res.* **39,** D830–D834.

Chmielarz, P. (2009). Cryopreservation of dormant European ash (*Fraxinus excelsior*) orthodox seeds. *Tree Physiol.* **29,** 1279–1285.

Choudhury, A. T. M. A., and Kennedy, I. R. (2004). Prospects and potentials for systems of biological nitrogen fixation in sustainable rice production. *Biol. Fertil. Soils* **39,** 219–227.

Christiansen, M. J., Andersen, S. B., and Ortiz, R. (2002). Diversity changes in an intensively bred wheat germplasm during the 20th century. *Mol. Breed.* **9,** 1–11.

Christie, M., Johnson, D., Stallings, C., and Hixon, M. (2010). Self recruitment and sweepstakes reproduction amid extensive gene flow in a coral-reef fish. *Mol. Ecol.* **19,** 1042–1057.

Clover, G., Hammons, S., and Unger, J.-G. (2010). International diagnostic protocols for regulated plant pests. *Bull. OEPP* **40,** 24–29.

Cnaani, A., and Levavi-Sivan, B. (2009). Sexual development in fish, practical applications for aquaculture. *Sex. Dev.* **3,** 164–175.

Cockett, N. E., McEwan, J. C., Dalrymple, B. P., Wu, C., Kijas, J., Maddox, J. F., Oddy, V. H., Nicholas, F., and Raadsma, H. (2010). Recent Advances In Sheep Genomics. The International Plant & Animal Genome XVIII Conference. January 9-13, San Diego, California.

Cockram, J., White, J., Zuluaga, D. L., Smith, D., Comadran, J., Macaulay, M., Luo, Z., Kearsey, M. J., Werner, P., Harrap, D., *et al.* (2010). Genome-wide association mapping to candidate polymorphism resolution in the unsequenced barley genome. *Proc. Natl. Acad. Sci. U.S.A.* **107,** 21611–21616.

Cohen, D., Bogeat-Triboulot, M. B., Tisserant, E., Balzergue, S., Martin-Magniette, M. L., Lelandais, G., Ningre, N., Renou, J. P., Tamby, J. P., Le Thiec, D., *et al.* (2010). Comparative transcriptomics of drought responses in *Populus*: A meta-analysis of genome-wide expression profiling in mature leaves and root apices across two genotypes. *BMC Genomics* **11,** 630.

Collado, M. C., Grzeskowiak, C., and Salminen, S. (2007). Probiotic strains and their combination inhibit in vitro adhesion of pathogens to pig intestinal mucosa. *Curr. Microbiol.* **55,** 260–265.

Collard, B. C., and Mackill, D. J. (2008). Marker-assisted selection: An approach for precision plant breeding in the twenty-first century. *Philos. Trans. R. Soc. Lond. B Biol. Sci.* **363,** 557–572.

Collins, N. C., Tardieu, F., and Tuberosa, R. (2008). QTL approaches for improving crop performance under abiotic stress conditions: Where do we stand? *Plant Physiol.* **147,** 469–486.

Comai, L., Young, K., Till, B. J., Reynolds, S. H., Greene, E. A., Codomo, C. A., Enns, L. C., Johnson, J. E., Burtner, C., Odden, A. R., *et al.* (2004). Efficient discovery of DNA polymorphisms in natural populations by Ecotilling. *Plant J.* **37,** 778–786.

Congiu, L., Pujolar, J. M., Forlani, A., Cenadelli, S., Dupanloup, I., Barbisan, F., Galli, A., and Fontana, F. (2011). Managing polyploidy in *ex situ* conservation genetics: The case of the critically endangered adriatic sturgeon (*Acipenser naccarii*). *PLoS One* **6,** e18249.

Cotter, D., O'Donovan, V., O'Maoiléidigh, N., Rogan, G., Roche, N., and Wilkins, N. P. (2000). An evaluation of the use of triploid Atlantic salmon (*Salmo salar* L.) in minimising the impact of escaped farmed salmon on wild populations. *Aquaculture* **186,** 61–75.

Cottrell, J. E., Krystufek, V., Tabbener, H. E., Milner, A. D., Connoly, T., Sing, L., Fluch, S., Burg, K., Lefevre, F., Achard, P., *et al.* (2005). Postglacial migration of *Populus nigra* L.: Lessons learnt from chloroplast DNA. *For. Ecol. Manage.* **219,** 293–312.

Critser, J. K., Agca, Y., and Gunasena, K. T. (1997). The cryobiology of mammalian oocytes. *In* "Reproductive Tissue Banking: Scientific Principle" (A. Karow and J. K. Critser, eds.), pp. 329–357. Academic Press, San Diego.

CropLife International (2006). Annual Report 2005/2006. http://www.croplife.org/files/documentspublished/1/en-us/PUB-AR/259_PUB-AR_2006_06_12_Annual_Report_2005-2006.pdf.

Crump, B. C., and Koch, E. W. (2008). Attached bacterial populations shared by four species of aquatic angiosperms. *Appl. Environ. Microbiol.* **74,** 5948–5957.

Cunningham, C. O. (2002). Molecular diagnosis of fish and shellfish diseases: Present status and potential use in disease control. *Aquaculture* **206,** 19–55.

Daetwyler, H. D., Villanueva, B., Bijma, P., and Woolliams, J. A. (2007). Inbreeding in genome-wide selection. *J. Anim. Breed. Genet.* **124,** 369–376.

Dalvit, C., De Marchi, M., and Cassandro, M. (2007). Genetic traceability of livestock products: A review. *Meat Sci.* **77,** 437–449.

D'Amato, A., Bachi, A., Fasoli, E., Boschetti, E., Peltre, G., Senechal, H., and Righetti, P. G. (2009). In-depth exploration of cow's whey proteome via combinatorial peptide ligand libraries. *J. Proteome Res.* **8,** 3925–3936.

Danan, S., Veyrieras, J. B., and Lefebvre, V. (2011). Construction of a potato consensus map and QTL meta-analysis offer new insights into the genetic architecture of late blight resistance and plant maturity traits. *BMC Plant Biol.* **11,** 16.

Danson, J., Mbogori, M., Kimani, M., Lagat, M., Kuria, A., and Diallo, A. (2006). Marker-assisted introgression of *opaque 2* gene into herbicide tolerant elite maize inbred lines. *Afr. J. Biotechnol.* **5,** 2417–2422.

Dar, W. D., Reddy, B. V. S., Gowda, C. L. L., and Ramesh, S. (2006). Genetic resources enhancement of ICRISAT-mandate crops. *Curr. Sci.* **91,** 880–884.

Dargie, J. D. (1990). Helping small farmers to improve their livestock. International Atomic Energy Agency, Vienna. *IAEA Yearbook* **1989,** B35–B55.

Das, P., Mukherjee, S., and Sen, R. (2008). Improved bioavailability and biodegradation of a model polyaromatic hydrocarbon by a biosurfactant producing bacterium of marine origin. *Chemosphere* **72,** 1229–1234.

Davidson, W. S., Koop, B. F., Jones, S. J. M., Iturra, P., Vidal, R., Maass, A., Jonassen, I., Lien, S., and Omholt, S. W. (2010). Sequencing the genome of the Atlantic salmon (*Salmo salar*). *Genome Biol.* **11,** 403.

Dawson, I. K., Hedley, P. E., Guarino, L., and Jaenicke, H. (2009). Does biotechnology have a role in the promotion of underutilised crops? *Food Policy* **34,** 319–328.

Day, J. G., Harding, K. C., Nadarajan, J., and Benson, E. E. (2008). Cryopreservation: Conservation of bioresources at ultra low temperatures. *In* "Molecular Biomethods Handbook" (J. M. Walker and R. Rapley, eds.), 2nd edn., pp. 917–947. Humana Press, Totowa.

Dekkers, J. C. M. (2004). Commercial application of marker- and gene-assisted selection in livestock: Strategies and lessons. *J. Anim. Sci.* **82,** E313–E328.

Demain, A. L. (2009). Biosolutions to the energy problem. *J. Ind. Microbiol. Biotechnol.* **36,** 319–332.

Demain, A. L., and Adrio, J. L. (2008). Strain improvement for production of pharmaceuticals and other microbial metabolites by fermentation. *Prog. Drug Res.* **65,** 252–289.

Deng, W., Xi, D., Mao, H., and Wanapat, M. (2008). The use of molecular techniques based on ribosomal RNA and DNA for rumen microbial ecosystem studies: A review. *Mol. Biol. Rep.* **35,** 265–274.

Denis, M., and Bouvet, J. M. (2011). Genomic selection in tree breeding: Testing accuracy of prediction models including dominance effect. *BMC Proc.* **5**(Suppl. 7), O13.

Denslow, N. D., Garcia-Reyero, N., and Barber, D. S. (2007). Fish 'n' chips: The use of microarrays for aquatic toxicology. *Mol. Biosyst.* **3,** 172–177.

De-Santis, C., and Jerry, D. R. (2007). Candidate growth genes in finfish—Where should we be looking? *Aquaculture* **272,** 22–38.

Deschamps, S., and Campbell, M. A. (2010). Utilization of next-generation sequencing platforms in plant genomics and genetic variant discovery. *Mol. Breed.* **25,** 553–570.

Deshmukh, R., Singh, A., Jain, N., Anand, S., Gacche, R., Singh, A., Gaikwad, K., Sharma, T., Mohapatra, T., and Singh, N. (2010). Identification of candidate genes for grain number in rice (*Oryza sativa* L.). *Funct. Integr. Genomics* **10,** 339–347.

Desnoyers, M., Giger-reverdin, S., Bertin, G., Duvaux-Ponter, C., and Sauvant, D. (2009). Metaanalysis of the influence of *Saccharomyces cerevisiae* supplementation on ruminal parameters and milk production of ruminants. *J. Dairy Sci.* **92,** 1620–1632.

Devey, M. E., Carson, S. D., Nolan, M. F., Matheson, A. C., Riini, C. T., and Hohepa, J. (2004). QTL associations for density and diameter in *Pinus radiata* and the potential for marker-aided selection. *Theor. Appl. Genet.* **108,** 516–524.

de Vicente, C., Metz, T., and Alercia, A. (2004). Descriptors for Genetic Markers Technologies IPGRI, Rome, Italy. http://www.bioversityinternational.org/index.php?id=19&user_bioversitypublications_pi1%5bshowUid%5d=2789.

de Vicente, M. C., Guzmán, F. A., Engels, J., and Rao, V. R. (2006). Genetic characterization and its use in decision-making for the conservation of crop germplasm. *In* "The Role of Biotechnology in Exploring and Protecting Agricultural Genetic Resources" (J. Ruane and A. Sonnino, eds.), pp. 151–172. FAO, Rome. ftp://ftp.fao.org/docrep/fao/009/a0399e/a0399e02.pdf.

Devlin, R. H., and Nagahama, Y. (2002). Sex determination and sex differentiation in fish: An overview of genetic, physiological, and environmental influences. *Aquaculture* **208,** 191–364.

DeWoody, J. A., and Avise, J. A. (2001). Genetic perspectives on natural history of fish mating systems. *J. Hered.* **92,** 167–172.

Dhali, A., Manik, R. S., Das, S. K., Singla, S. K., and Palta, P. (2000). Effect of ethylene glycol concentration and exposure time on post-vitrification survival and *in vitro* maturation rate of buffalo oocytes. *Theriogenolgy* **50,** 521–530.

Diagne, A., Midingoyi, G. S., and Kinkingninhoun, F. (2009). The impact of NERICA adoption on rice yield in West Africa: Evidence from four countries. Paper presented at the International Association of Agricultural Economists 2009 Conference, Beijing, China, August 16-22, 2009.

Dick, C. W., Hardy, O. J., Jones, F. A., and Petit, R. J. (2008). Spatial scales of pollen and seed mediated gene flow in tropical rain forest trees. *Trop. Plant Biol.* **1,** 20–33.

Dick, C. W., and Kress, J. W. (2009). Dissecting tropical plant diversity with forest plots and a molecular toolkit. *Bioscience* **59,** 745–755.

Dillon, S. K., Nolan, M. F., Wu, H., and Southerton, S. G. (2010). Association genetics reveal candidate gene SNPs affecting wood properties in *Pinus radiata. Aust. For.* **73,** 185–190.

Ding, L., Zhou, Q., Wang, L., and Zhang, Q. (2011). Dynamics of bacterial community structure in a full-scale wastewater treatment plant with anoxic-oxic configuration using 16S rDNA PCR-DGGE fingerprints. *Afr. J. Biotechnol.* **10,** 589–600.

Diwan, A. D., Ayyappan, S., Lal, K. K., and Lakra, W. S. (2010). Cryopreservation of fish gametes and embryos. *Indian J. Anim. Sci.* **80**(Suppl. 1), 109–124.

Djordjevic, B., Škugor, S., Jørgensen, S. M., Øverland, M., Mydland, L. T., and Krasnov, A. (2009). Modulation of splenic immune responses to bacterial lipopolysaccharide in rainbow trout (*Oncorhynchus mykiss*) fed lentinan, a beta-glucan from mushroom *Lentinula edodes. Fish Shellfish Immunol.* **26,** 201–209.

Dodd, D., Moon, Y.-H., Swaminathan, K., Mackie, R. I., and Cann, I. K. O. (2010). Transcriptomic analyses of xylan degradation by *Prevotella bryantii* and insights into energy acquisition by xylanolytic bacteroidetes. *J. Biol. Chem.* **285,** 30261–30273.

Domenech, J., Lubroth, J., Eddi, C., Martin, V., and Roger, F. (2006). Regional and international approaches on prevention and control of animal transboundary and emerging diseases. *Ann. N. Y. Acad. Sci.* **1081,** 90–107.

Dong, Q. X., Eudelie, B., Huang, C. J., Allen, S. K., and Tiersch, T. R. (2005). Commercial-scale sperm cryopreservation of diploid and tetraploid oysters, *Crassostrea gigas. Cryobiology* **50,** 1–16.

Donini, P., Law, J. R., Koebner, R. M. D., Reeves, J. C., and Cooke, R. J. (2000). Temporal trends in the diversity of UK wheat. *Theor. Appl. Genet.* **100,** 912–917.

Douglas, C. J., and DiFazio, S. P. (2010). The *Populus* genome and comparative genomics. *In* "Genetics and Genomics of *Populus*" (S. Jansson, R. Bhalerao, and A. Groover, eds.), Plant Genetics and Genomics: Crops and Models, Vol. 8, pp. 67–90. Springer, New York, Dordrecht, Heidelberg, London.

Douglas, S. E., Knickle, L. C., Williams, J., Flight, R. M., and Reith, M. E. (2008). A first generation Atlantic halibut *Hippoglossus hippoglossus* (L.) microarray: Application to developmental studies. *J. Fish Biol.* **72,** 2391–2406.

Dow, B. D., and Ashley, M. V. (1998). High levels of gene flow in bur oak revealed by paternity analysis using microsatellites. *J. Hered.* **89,** 62–70.

Dreher, K., Khairallah, M., Ribaut, J. M., and Morris, M. (2003). Money matters (I): Costs of field and laboratory procedures associated with conventional and marker-assisted maize breeding at CIMMYT. *Mol. Breed.* **11,** 221–234.

Dulloo, E. (2011). Complementary conservation actions. *In* "Crop Wild relatives. A Manual of *In Situ* Conservation" (D. Hunter and V. H. Heywood, eds.), pp. 275–294. Routledge, London.

Dulloo, M. E., Ramanatha Rao, V., Engelmann, F., and Engels, J. (2005). Complementary conservation of coconuts. *In* "Coconut Genetic Resources" (P. Batugal, V. R. Rao, and J. Oliver, eds.), pp. 75–90. IPGRI-APO, Serdang.

Dulloo, M. E., Ebert, A. W., Dussert, S., Gotor, E., Astorga, C., Vasquez, N., Rakotomalala, J. J., Rabemiafara, A., Eira, M., Bellachew, B., *et al.* (2009). Cost efficiency of cryopreservation as a long-term conservation method for coffee genetic resources. *Crop Sci.* **49,** 2123–2138.

Dulloo, M. E., Hunter, D., and Borelli, T. (2010). *Ex situ* and *in situ* conservation of agricultural biodiversity: Major advances and research needs. *Not. Bot. Hort. Agrobot. Cluj* **38,** 123–135.

Dumpala, P. R., Gülsoy, N., Lawrence, M. L., and Karsi, A. (2010). Proteomic analysis of the fish pathogen *Flavobacterium columnare*. *Proteome Sci.* **8,** 26.

Dunham, R. A. (2003). Status of genetically modified (transgenic) fish: Research and application. *Working Paper Topic 2*. Food and Agriculture Organization/World Health Organization expert hearings on biotechnology and food safety. ftp://ftp.fao.org/es/esn/food/GMtopic2.pdf.

Dunham, R. A. (2004). Aquaculture and Fisheries Biotechnology: Genetic Approaches. CABI publishing, Oxfordshire.

Dunwell, J. M. (2010). Haploids in flowering plants: Origins and exploitation. *Plant Biotechnol. J.* **8,** 377–424.

Duponnois, R., Kisa, M., Prin, Y., Ducousso, M., Plenchette, C., Lepage, M., and Galiana, A. (2008). Soil factors influencing the growth response of *Acacia holosericea* A Cunn. ex G. Don to ectomycorrhizal inoculation. *New Forests* **35,** 105–117.

Duran, C., Edwards, D., and Batley, J. (2008). Genetic maps and the use of synteny. *Methods Mol. Biol.* **513,** 41–55.

Duran, C., Appleby, N., Edwards, D., and Batley, J. (2009). Molecular genetic markers: Discovery, applications, data storage and visualisation. *Curr. Bioinform.* **4,** 16–27.

Durkovic, J., and Misalova, A. (2008). Micropropagation of temperate noble hardwoods: An overview. *Funct. Plant Sci. Biotechnol.* **2,** 1–19.

Dussert, S., and Engelmann, F. (2006). New determinants of coffee (*Coffea arabica* L.) seed tolerance to liquid nitrogen exposure. *Cryo Letters* **27,** 169–178.

Dussert, S., Chabrillange, N., Anthony, F., Engelmann, F., Recalt, C., and Hamon, S. (1997). Variability in storage response within a coffee (*Coffea* spp.) core collection under slow growth conditions. *Plant Cell Rep.* **16,** 344–348.

Dwivedi, S. L., Crouch, J. H., Mackill, D. J., Xu, Y., Blair, M. W., Ragot, M., Upadhyaya, H. D., and Ortiz, R. (2007). The molecularization of public sector crop breeding: Progress, problems, and prospects. *Adv. Agron.* **95,** 163–318.

Eathington, S. R., Crosbie, T. M., Edwards, M. D., Reiter, R. S., and Bull, J. K. (2007). Molecular markers in a commercial breeding program. *Crop Sci.* **47,** S154–S163.

Ebana, K., Kojima, Y., Fukuoka, S., Nagamine, T., and Kawase, M. (2008). Development of mini core collection of Japanese rice landrace. *Breed. Sci.* **58,** 281–291.

Eckert, A. J., Pande, B., Ersoz, E. S., Wright, M. H., Rashbrook, V. K., Nicolet, C. M., and Neale, D. B. (2009a). High-throughput genotyping and mapping of single nucleotide polymorphisms in loblolly pine (*Pinus taeda* L.). *Tree Genet. Genomes* **5,** 225–234.

Eckert, A. J., Bower, A. D., Wegrzyn, J. L., Pande, B., Jermstad, K. D., Krutovsky, K. V., St Clair, J. B., and Neale, D. B. (2009b). Association genetics of coastal Douglas fir (*Pseudotsuga menziesii* var. menziesii, Pinaceae). I. Cold-hardiness related traits. *Genetics* **182,** 1289–1302.

Egito, A. A., Paiva, S. R., Albuquerque, M. S. M., Mariante, A. S., Almeida, L. D., Castro, S. R., and Grattapaglia, D. (2007). Microsatellite based genetic diversity and relationships among ten Creole and commercial cattle breeds raised in Brazil. *BMC Genet.* **8,** 83–96.

Eglinton, J., Coventry, S., and Chalmers, K. (2006). Breeding outcomes from molecular genetics. *In* "Breeding for Success: Diversity in Action" (C. F. Mercer, ed.), Proceedings of the 13th Australasian Plant Breeding Conference, Christchurch, New Zealand. pp. 743–749.

Ejeta, G. (2007). Breeding for Striga resistance in sorghum: Exploitation of an intricate host–parasite biology. *Crop Sci.* **47,** S216–S227.

Ejeta, G., and Gressel, J. (2007). Integrating New Technologies for Striga Control. Towards Ending the Witch-Hunt. World Scientific Publishing, Singapore.

Eldridge, W. H., and Killebrew, K. (2008). Genetic diversity over multiple generations of supplementation: An example from Chinook salmon using microsatellite and demographic data. *Conserv. Genet.* **9,** 13–28.

El-Kassaby, Y. A., and Lstibůrek, M. (2009). Breeding without breeding. *Genet. Res.* **91,** 111–120.

El-Kassaby, Y. A., Cappa, E. P., Liewlaksaneeyanawin, C., Klápště, J., and Lstibůrek, M. (2011). Breeding without breeding: Is a complete pedigree necessary for efficient breeding? *PLoS One* **6,** e25737.

Ellstrand, N., Prentice, H., and Hancock, J. (1999). Gene flow and introgression from domesticated plants into their wild relatives. *Annu. Rev. Ecol. Syst.* **30,** 539–563.

Engelmann, F. (2011). Use of biotechnologies for the conservation of plant biodiversity. *In Vitro Cell. Dev. Biol. Plant* **47,** 5–16.

Engels, J. M. M., Maggioni, L., Maxted, N., and Dulloo, M. E. (2008). Complementing *in situ* conservation with *ex situ* measures. *In* "Conserving Plant Genetic Diversity in Protected Areas: Population Management of Crop Wild Relatives" (J. M. Iriondo, N. Maxted, and M. E. Dulloo, eds.), pp. 169–181. CABI publishing, Oxfordshire.

Ewart, K. V., Williams, J., Richards, R. C., Gallant, J. W., Melville, K., and Douglas, S. E. (2008). The early response of Atlantic salmon (*Salmo salar*) macrophages exposed *in vitro* to *Aeromonas salmonicida* cultured in broth and in fish. *Dev. Comp. Immunol.* **32,** 380–390.

Ezaz, M. T., Myers, J. M., Powell, S. F., McAndrew, B. J., and Penman, D. J. (2004). Sex ratios in the progeny of androgenetic and gynogenetic YY male Nile tilapia, *Oreochromis niloticus* L. *Aquaculture* **232,** 205–214.

Fachinger, V., Bischoff, R., Jedidia, S. B., Saalmüller, A., and Elbers, K. (2008). The effect of vaccination against porcine circovirus type 2 in pigs suffering from porcine respiratory disease complex. *Vaccine* **26,** 1488–1499.

Fan, B., Du, Z.-H., Gorbach, D. M., and Rothschild, M. F. (2010). Development and application of high-density SNP arrays in genomic studies of domestic animals. *Asian-Aust. J. Anim. Sci.* **23,** 833–847.

Fan, B., Gorbach, D. M., and Rothschild, M. F. (2011). The pig genome project has plenty to squeal about. *Cytogenet. Genome Res.* **134,** 9–18.

FAO (1994). Biotechnology in forest tree improvement with special references to developing countries. *Forestry Paper No. 118.* http://www.fao.org/DOCREP/006/T2114E/T2114E00.HTM.

FAO (1997). The state of the world's plant genetic resources for food and agriculture. http://apps3.fao.org/wiews/docs/swrfull.pdf.

FAO (1998). Food quality and safety systems—A training manual on food hygiene and the hazard analysis and critical control point (HACCP) system. FAO Agricultural Policy and Economic Development Series. http://www.fao.org/docrep/W8088E/W8088E00.htm.

FAO (1999). The global strategy for the management of farm animal genetic resources: Executive brief. http://lprdad.fao.org/cgi-bin/getblob.cgi?sid=-1,50006152.

FAO (2001). Glossary of biotechnology for food and agriculture—A revised and augmented edition of the glossary of biotechnology and genetic engineering. By A. Zaid, H. G. Hughes, E. Porceddu, and F. Nicholas. *FAO Research and Technology Paper* 9. http://www.fao.org/docrep/004/y2775e/y2775e00.htm.

FAO (2002). Gene flow from GM to non-GM populations in the crop, forestry, animal and fishery sectors. *Background Document to Conference 7 of the FAO Biotechnology Forum (31 May to 6 July 2002)*. http://www.fao.org/biotech/biotech-forum/conference-7/en/.

FAO (2003). Molecular marker assisted selection as a potential tool for genetic improvement of crops, forest trees, livestock and fish in developing countries. *Background Document to Conference 10 of the FAO Biotechnology Forum (17 November to 14 December 2003)*. http://www.fao.org/biotech/biotech-forum/conference-10/en/.

FAO (2004). Preliminary review of biotechnology in forestry, including genetic modification. *Forest Genetic Resources Working Paper FGR/59E*. Rome. www.fao.org/docrep/008/ae574e/ae574e00.htm.

FAO (2005). Status of research and application of crop biotechnologies in developing countries—Preliminary assessment. By Z. Dhlamini, C. Spillane, J. P. Moss, J. Ruane, N. Urquia, and A. Sonnino. Rome. www.fao.org/docrep/008/y5800e/y5800e00.htm.

FAO (2006). FAO/WHO guidance to governments on the application of HACCP in small and/or less-developed food businesses. *FAO Food and Nutrition Paper* 86. Rome. www.fao.org/docrep/009/a0799e/a0799e00.htm.

FAO (2007a). *In* "Marker-assisted Selection: Current Status and Future Perspectives in Crops, Livestock, Forestry and Fish" (E. Guimarães, J. Ruane, B. Scherf, A. Sonnino, and J. Dargie, eds.). FAO, Rome. http://www.fao.org/docrep/010/a1120e/a1120e00.htm.

FAO (2007b). The state of the world's animal genetic resources for food and agriculture. (B. Rischkowsky and D. Pilling, eds.). FAO, Rome. http://www.fao.org/docrep/010/a1250e/a1250e00.htm.

FAO (2008). Aquaculture development. 3. Genetic resource management. *FAO Technical Guidelines for Responsible Fisheries*. No. 5, Suppl. 3. Rome. 125p. http://www.fao.org/docrep/011/i0283e/i0283e00.htm.

FAO (2009c). Learning from the past: Successes and failures with agricultural biotechnologies in developing countries over the last 20 years. *Background Document to Conference 16 of the FAO Biotechnology Forum (8 June to 8 July 2009)*. http://www.fao.org/fileadmin/user_upload/abdc/documents/emailconf.pdf.

FAO (2009a). How to feed the world in 2050. Paper presented at the High Level Expert Forum, Rome 12–13 October 2009. Rome. http://www.fao.org/fileadmin/templates/wsfs/docs/expert_paper/How_to_Feed_the_World_in_2050.pdf.

FAO (2009b). The State of Food and Agriculture. Livestock in the Balance. FAO, Rome. http://www.fao.org/docrep/012/i0680e/i0680e.pdf.

FAO (2010a). The State of World Fisheries and Aquaculture 2010. FAO, Rome. http://www.fao.org/docrep/013/i1820e/i1820e00.htm.

FAO (2010b). Global Forest Resources Assessment 2010. Main Report. FAO, Rome. http://foris.fao.org/static/data/fra2010/FRA2010_Report_en_WEB.pdf.

FAO (2010c). *The Second Report on the State of the World's Plant Genetic Resources for Food and Agriculture*. FAO, Rome. http://www.fao.org/docrep/013/i1500e/i1500e00.htm.

FAO (2010d). FAO Yearbook. Fishery and Aquaculture Statistics 2008. FAO, Rome. http://www.fao.org/docrep/013/i1890t/i1890t.pdf.

FAO (2011a). Current status and options for crop biotechnologies in developing countries. Biotechnologies for Agricultural Development: Proceedings of the FAO International Technical Conference on Agricultural Biotechnologies in Developing Countries: Options and Opportunities in Crops, Forestry, Livestock, Fisheries and Agro-industry to Face the Challenges of Food Insecurity and Climate Change (ABDC-10). pp. 2–77. FAO, Rome.

FAO (2011b). Current status and options for forest biotechnologies in developing countries. Biotechnologies for Agricultural Development: Proceedings of the FAO International Technical Conference on Agricultural Biotechnologies in Developing Countries: Options and Opportunities in Crops, Forestry, Livestock, Fisheries and Agro-industry to Face the Challenges of Food Insecurity and Climate Change (ABDC-10). pp. 78–122. FAO, Rome.

FAO (2011c). Current status and options for livestock biotechnologies in developing countries. Biotechnologies for Agricultural Development: Proceedings of the FAO International Technical Conference on Agricultural Biotechnologies in Developing Countries: Options and Opportunities in Crops, Forestry, Livestock, Fisheries and Agro-industry to Face the Challenges of Food Insecurity and Climate Change (ABDC-10). pp. 123–190. FAO, Rome.

FAO (2011d). Current status and options for biotechnologies in aquaculture and fisheries in developing countries. Biotechnologies for Agricultural Development: Proceedings of the FAO International Technical Conference on Agricultural Biotechnologies in Developing Countries: Options and Opportunities in Crops, Forestry, Livestock, Fisheries and Agro-industry to Face the Challenges of Food Insecurity and Climate Change (ABDC-10). pp. 191–239. FAO, Rome.

FAO (2011e). Current status and options for biotechnologies in food processing and in food safety in developing countries. Biotechnologies for Agricultural Development: Proceedings of the FAO International Technical Conference on Agricultural biotechnologies in Developing Countries: Options and Opportunities in Crops, Forestry, Livestock, Fisheries and Agro-industry to Face the Challenges of Food Insecurity and Climate Change (ABDC-10). pp. 240–277. FAO, Rome.

FAO (2011f). Targeting agricultural biotechnologies to the poor. Biotechnologies for Agricultural Development: Proceedings of the FAO International Technical Conference on Agricultural Biotechnologies in Developing Countries: Options and Opportunities in Crops, Forestry, Livestock, Fisheries and Agro-industry to Face the Challenges of Food Insecurity and Climate Change (ABDC-10). pp. 328–370. FAO, Rome.

FAO (2011g). Enabling R&D for agricultural biotechnologies. Biotechnologies for Agricultural Development: Proceedings of the FAO International Technical Conference on Agricultural Biotechnologies in Developing Countries: Options and Opportunities in Crops, Forestry, Livestock, Fisheries and Agro-industry to Face the Challenges of Food Insecurity and Climate Change (ABDC-10). pp. 371–419. FAO, Rome.

FAO (2011h). Ensuring access to the benefits of R&D. Biotechnologies for Agricultural Development: Proceedings of the FAO International Technical Conference on Agricultural Biotechnologies in Developing Countries: Options and Opportunities in Crops, Forestry, Livestock, Fisheries and Agro-industry to Face the Challenges of Food Insecurity and Climate Change (ABDC-10). pp. 420–466. FAO, Rome.

FAO/IAEA (2008). Atoms for food: A global partnership. http://www.iaea.or.at/Publications/Booklets/Fao/fao1008.pdf.

FAO/NACA (2000). Asia regional technical guidelines on health management for the responsible movement of live aquatic animals and the Beijing consensus and implementation strategy. *FAO Fisheries Technical Paper No. 402*. Rome. www.fao.org/docrep/005/x8485e/x8485e00.htm.

FAO/OIE/WHO (2006). *Report of a joint FAO/OIE/WHO expert consultation on antimicrobial use in aquaculture and antimicrobial resistance*. Seoul, Republic of Korea, 13–16 June 2006. ftp://ftp.fao.org/ag/agn/food/aquaculture_rep_13_16june2006.pdf.

FAO/WHO (2001). *Report of a joint FAO/WHO expert consultation on evaluation of health and nutritional properties of probiotics in food including powder milk with live lactic acid bacteria*. Córdoba, Argentina, 1–4 October 2001. http://www.who.int/foodsafety/publications/fs_management/en/probiotics.pdf.

FAO/WHO (2006). Probiotics in food. Health and nutrional properties and guidelines for evaluation. *FAO Food and Nutrition Paper* 85. ftp://ftp.fao.org/docrep/fao/009/a0512e/a0512e00.pdf.

Faramarzi, M., Jafaryan, H., Farahi, A., Boloki, M. L., and Iranshahi, F. (2011). The effects on growth and survival of probiotic *Bacillus* spp. fed to Persian sturgeon (*Acipencer persicus*) larvae. *AACL Bioflux* **4,** 10–14.

Farzanfar, A. (2006). The use of probiotics in shrimp aquaculture. *FEMS Immunol. Med. Microbiol.* **48,** 149–158.

Fears, R. (2007). Genomics and genetic resources for food and agriculture. *Background Study Paper No. 34*. ftp://ftp.fao.org/docrep/fao/meeting/014/k0174e.pdf.

Feindel, N. J., Benfey, T. J., and Trippel, E. A. (2010). Competitive spawning success and fertility of triploid male Atlantic cod *Gadus morhua*. *Aquacult. Environ. Interact.* **1,** 47–55.

Feldmann, U., Dyck, V., Mattioli, R., and Jannin, J. (2005). Potential impact of tsetse fly control involving the sterile insect technique. *In* "Sterile Insect Technique. Principles and Practice in Area-Wide Integrated Pest Management" (V. A. Dyck, J. Hendrichs, and A. S. Robinson, eds.), pp. 701–723. Springer, Netherlands.

Feng, C., Yin, Z., Ma, Y., Zhang, Z., Chen, L., Wang, B., Li, B., Huang, Y., and Wang, Q. (2011). Cryopreservation of sweetpotato (*Ipomoea batatas*) and its pathogen eradication by cryotherapy. *Biotechnol. Adv.* **29,** 84–93.

Fernández-Luqueño, F., Valenzuela-Encinas, C., Marsch, R., Martínez-Suárez, C., Vázquez-Núñez, E., and Dendooven, L. (2011). Microbial communities to mitigate contamination of PAHs in soil—possibilities and challenges: A review. *Environ. Sci. Pollut. Res. Int.* **18,** 12–30.

Fernández-Manjarrés, J. F., Gerard, P. R., Dufour, J., Raquin, C., and Frascaria-Lacoste, N. (2006). Differential patterns of morphological and molecular hybridization between *Fraxinus excelsior* L. and *Fraxinus angustifolia* Vahl (Oleaceae) in eastern and western France. *Mol. Ecol.* **15,** 3245–3257.

Fernie, A. R., and Schauer, N. (2009). Metabolomics-assisted breeding: A viable option for crop improvement? *Trends Genet.* **25,** 39–48.

Ferraresso, S., Vitulo, N., Mininni, A. N., Romualdi, C., Cardazzo, B., Negrisolo, E., Reinhardt, R., Canario, A. V. M., Patarnello, T., and Bargelloni, L. (2008). Development and validation of a gene expression oligo microarray for the gilthead sea bream (*Sparus aurata*). *BMC Genomics* **9,** 580.

Ferraresso, S., Milan, M., Pellizzari, C., Vitulo, N., Reinhardt, R., Canario, A. V. M., Patarnello, T., and Bargelloni, L. (2010). Development of an oligo DNA microarray for the European sea bass and its application to expression profiling of jaw deformity. *BMC Genomics* **11,** 354.

Finkeldey, R., Leinemann, L., and Gailing, O. (2010). Molecular genetic tools to infer the origin of forest plants and wood. *Appl. Microbiol. Biotechnol.* **85,** 1251–1258.

Fitton, N., and Thomas, J. S. (2009). Gastrointestinal dysfunction. *Surgery* **27,** 492–495.

Flajshans, M., Gela, D., Kocour, M., Buchtova, H., Rodina, M., Psenicka, M., Kaspar, V., Piackova, V., Sudova, E., and Linhart, O. (2010). A review on the potential of triploid tench for aquaculture. *Rev. Fish Biol. Fisheries* **20,** 317–329.

Flannery, B. G., Beacham, T. D., Candy, J. R., Holder, R. R., Maschmann, G. F., Kretschmer, E. J., and Wenburg, J. K. (2010). Mixed-stock analysis of Yukon River chum salmon: Application and validation in a complex fishery. *N. Am. J. Fish. Manag.* **30,** 1324–1338.

Flavell, R. (2008). Role of model plant species. *Methods Mol. Biol.* **513,** 1–18.

Flint, A. P. F., and Woolliams, J. A. (2008). Precision animal breeding. *Philos. Trans. R. Soc. B* **363,** 573–590.

Flint, J. F., and Garner, M. R. (2009). Feeding beneficial bacteria: A natural solution for increasing efficiency and decreasing pathogens in animal agriculture. *J. Appl. Poult. Res.* **18,** 367–378.

Flores, M., and Toldra, F. (2011). Microbial enzymatic activities for improved fermented meats. *Trends Food Sci. Technol.* **22,** 81–90.

Flores-Valverde, A. M., Horwood, J., and Hill, E. M. (2010). Disruption of the steroid metabolome in fish caused by exposure to the environmental estrogen 17α-ethinylestradiol. *Environ. Sci. Technol.* **44,** 3552–3558.

Flury, C., Tapio, M., Sonstegard, T., Drögemüller, C., Leeb, T., Simianer, H., Hanotte, O., and Rieder, S. (2010). Effective population size of an indigenous Swiss cattle breed estimated from linkage disequilibrium. *J. Anim. Breed. Genet.* **127,** 339–347.

Fogarty, N. M., Maxwell, W. M. C., Eppleston, J., and Evans, G. (2000). The viability of transferred sheep embryos after long-term cryopreservation. *Reprod. Fertil. Dev.* **12,** 31–37.

Forne, I., Abian, J., and Cerda, J. (2010). Fish proteome analysis: Model organisms and nonsequenced species. *Proteomics* **10,** 858–872.

Franco, J., Crossa, J., Warburton, M. L., and Taba, S. (2006). Sampling strategies for conserving maize diversity when forming core subsets using genetic markers. *Crop Sci.* **46,** 854–864.

Frankham, R., Ballou, J. D., Eldridge, M. D., Lacy, R. C., Ralls, K., Dudash, M. R., and Fenster, C. B. (2011). Predicting the probability of outbreeding depression. *Conserv. Biol.* **25,** 465–475.

Freeman, A. R., Bradley, D. G., Nagda, S., Gibson, J. P., and Hanotte, O. (2006). Combination of multiple microsatellite data sets to investigate genetic diversity and admixture of domestic cattle. *Anim. Genet.* **37,** 1–9.

Freitas, P. D., Calgaro, M. R., and Galetti, P. M., Jr. (2007). Genetic diversity within and between broodstocks of the white shrimp *Litopenaeus vannamei* (Boone, 1931) (Decapoda, Penaeidae) and its implication for the gene pool conservation. *Braz. J. Biol.* **67**(Suppl. 4), 939–943.

Frizzo, L. S., Sotto, L. P., Zbrun, M. V., Bertozzi, E., Sequeira, G., Armesto, R. R., and Rosmini, M. R. (2010). Lactic acid bacteria to improve growth performance in young calves fed milk replacer and spray-dried whey powder. *Anim. Feed Sci. Technol.* **157,** 159–167.

Fu, Y., Peterson, G. W., Richards, K. W., Tarn, T. R., and Percy, J. E. (2009). Genetic diversity of Canadian and exotic potato germplasm revealed by simple sequence repeat markers. *Am. J. Potato Res.* **86,** 38–48.

Fuchs, E. J., and Hamrick, J. L. (2010). Genetic diversity in the endangered tropical tree, *Guaiacum sanctum* (Zygophyllaceae). *J. Hered.* **101,** 284–291.

Fuglie, K.O, Zhang, L., Salazar, L.F, and Walker, T. (1999). Economic impact of virus free sweet-potato seed in Shandong Province, China. Lima, Peru, International Potato Center. www.eseap.cipotato.org/MF-ESEAP/Fl-Library/Eco-Imp-SP.pdf.

Fuji, K., Hasegawa, O., Honda, K., Kumasaka, K., Sakamoto, T., and Okamoto, N. (2007). Marker-assisted breeding of a lymphocystis disease-resistant Japanese flounder (*Paralichthys olivaceus*). *Aquaculture* **272,** 291–295.

Gaggìa, F., Mattarelli, P., and Biavati, B. (2010). Probiotics and prebiotics in animal feeding for safe food production. *Int. J. Food Microbiol.* **141**(Suppl. 1), S15–S28.

Ganal, M. W., Altmann, T., and Roder, M. (2009). SNP identification in crop plants. *Curr. Opin. Plant Biol.* **12,** 211–217.

Gandini, G., Pizzi, F., Stella, A., and Boettcher, P. J. (2007). The costs of breed reconstruction from cryopreserved material in mammalian livestock species. *Genet. Sel. Evol.* **39,** 465–479.

Ganeshan, S., and Rajashekaran, P. E. (2000). Current status of pollen cryopreservation research: Relevance to tropical agriculture. *In* "Cryopreservation of Tropical Plant Germplasm—Current Research Progress and Applications" (F. Engelmann and H. Takagi, eds.), pp. 360–365. JIRCAS, Tsukuba.

Gangopadhyay, G., Gangopadhyay, S. B., Poddar, R., Gupta, S., and Mukherjee, K. K. (2003). Micropropagation of *Tectona grandis*: Assessment of genetic fidelity. *Biol. Plantarum* **46,** 459–461.

Gänzle, M. G. (2009). From gene to function: Metabolic traits of starter cultures for improved quality of cereal foods. *Int. J. Food Microbiol.* **134,** 29–36.

Gao, D., and Tao, Y. (2012). Current molecular biologic techniques for characterizing environmental microbial community. *Front. Environ. Sci. Eng. China* **6,** 82–97.

Gao, Y., Zhang, Y. H., Jiang, H., Xiao, S. Q., Wang, S., Ma, Q., Sun, G. J., Li, F. J., Deng, Q., Dai, L. S., *et al.* (2011). Detection of differentially expressed genes in the longissimus dorsi of Northeastern Indigenous and Large White pigs. *Genet. Mol. Res.* **10,** 779–791.

Gardner, S. N., and Slezak, T. (2010). Scalable SNP analyses of 100+ bacterial or viral genomes. *J. Forensic Res.* **1,** 107–111.

Garg, L., Bhandari, N. N., Rani, V., and Bhojwani, S. S. (1996). Somatic embryogenesis and regeneration of triploid plants in endosperm cultures of *Acacia nilotica*. *Plant Cell Rep.* **15,** 855–858.

Gehring, A. G. (2011). High-throughput biosensors for multiplexed foodborne pathogen detection. *Annu. Rev. Anal. Chem. (Palo Alto Calif.)* **4,** 151–172.

Gerdts, V., Mutwiri, G. K., Tikoo, S. K., and Babiuk, L. A. (2006). Mucosal delivery of vaccines in domestic animals. *Vet. Res.* **37,** 487–510.

Germana, M. A. (2011). Gametic embryogenesis and haploid technology as valuable support to plant breeding. *Plant Cell Rep.* **30,** 839–857.

Ghazanfar, S., and Azim, A. (2009). Metagenomics and its application in rumen ecosystem: Potential biotechnological prospects. *Pak. J. Nutr.* **8,** 1309–1315.

Ghazanfar, S., Azim, A., Ghazanfar, M. A., Anjum, M. I., and Begum, I. (2010). Metagenomics and its application in soil microbial community studies: Biotechnological prospects. *J. Anim. Plant Sci.* **6,** 611–622.

Ghislain, M., Andrade, D., Rodríguez, F., Hijmans, R. J., and Spooner, D. M. (2006). Genetic analysis of the cultivated potato *Solanum tuberosum* L. Phureja Group using RAPDs and nuclear SSRs. *Theor. Appl. Genet.* **113,** 1515–1527.

Gilchrist, E., and Haughn, G. (2010). Reverse genetics techniques: Engineering loss and gain of gene function in plants. *Brief. Funct. Genomics* **9,** 103–110.

Gilchrist, E. J., Haughn, G. W., Ying, C. C., Otto, S. P., Zhuang, J., Cheung, D., Hamberger, B., Aboutorabi, F., Kalynyak, T., Johnson, L., *et al.* (2006). Use of EcoTILLING as an efficient SNP discovery tool to survey genetic variation in wild populations of *Populus trichocarpa*. *Mol. Ecol.* **15,** 1367–1378.

Glocke, P., Delaporte, K., Collins, G., and Sedgley, M. (2006). Micropropagation of juvenile tissue of *Eucalyptus erythronema* × *Eucalyptus stricklandii* cv. 'urrbrae gem'. *In Vitro Cell. Dev. Biol. Plant* **42,** 139–143.

Glover, K. A. (2010). Forensic identification of fish farm escapees: The Norwegian experience. *Aquaculture Environ. Interact.* **1,** 1–10.

Glowatzki-Mullis, M. L., Gaillard, C., Wigger, G., and Fries, R. (1995). Microsatellite-based parentage control in cattle. *Anim. Genet.* **26,** 7–12.

Glynn, B., Lahiff, S., Wernecke, M., Barry, T., Smith, T. J., and Maher, M. (2006). Current and emerging molecular diagnostic technologies applicable to bacterial food safety. *Int. J. Dairy Technol.* **59,** 126–139.

Goh, D. K. S., Chaix, G., Baillères, H., and Monteuuis, O. (2007). Mass production and quality control of teak clones for tropical plantations: The Yayasan Sabah Group and Forestry Department of CIRAD joint project as a case study. *Bois et Forêts des Tropiques* **33,** 6–9.

Goh, D. K. S., Chang, F., Jilimin, M., and Japarudin, Y. (2010). Tissue culture propagation and dispatch of quality teak clones. *AsPac J. Mol. Biol. Biotechnol.* **18,** 147–149.

Gomez-Pando, L. R., Jimenez-Davalos, J., Eguiluz-de la Barra, A., Aguilar-Castellanos, E., Falconí-Palomino, J., Ibañez-Tremolada, M., Varela, M., and Lorenzo, J. C. (2009). Field performance of new *in vitro* androgenesis-derived double haploids of barley. *Euphytica* **166,** 269–276.

Gomez-Uchida, D., Seeb, J. E., Smith, M. J., Habicht, C., Quinn, T. P., and Seeb, L. W. (2011). Single nucleotide polymorphisms unravel hierarchical divergence and signatures of selection among Alaskan sockeye salmon (*Oncorhynchus nerka*) populations. *BMC Evol. Biol.* **11,** 48.

Gonzalez-Arnao, M. T., Panta, A., Roca, W. M., Escobar, R. H., and Engelmann, F. (2008). Development and large scale application of cryopreservation techniques for shoot and somatic embryo cultures of tropical crops. *Plant Cell Tissue Organ Cult.* **92,** 1–13.

González-Fernández, R., Prats, E., and Jorrín-Novo, J. V. (2010). Proteomics of plant pathogenic fungi. *J. Biomed. Biotechnol.* **2010,** 932527.

Gootwine, E., Zenu, A., Bor, A., Yossafi, S., Rosov, A., and Pollott, G. E. (2001). Genetic and economic analysis of introgression the B allele of the *FecB* (Booroola) gene into the Awassi and Assaf dairy breeds. *Livest. Prod. Sci.* **71,** 49–58.

Gootwine, E., Rozov, A., Bor, A., and Reicher, S. (2003). Effects of the *FecB* (Booroola) gene on reproductive and productive traits in the Assaf breed. Proceedings of the International Workshop on Major Genes and QTL in Sheep and Goats. CD-ROM communication no. 2-12, 4. Toulouse.

Govindarajan, M., Balandreau, J., Muthukumarasamy, R., Kwon, S.-W., Weon, H.-Y., and Lakshminarasimhan, C. (2008). Effects of the inoculation of *Burkholderia vietnamiensis* and related endophytic diazotrophic bacteria on grain yield of rice. *Microb. Ecol.* **55,** 21–37.

Goyache, F., Álvarez, I., Fernández, I., Pérez-Pardal, L., Royo, L. J., and Lorenzo, L. (2011). Usefulness of molecular-based methods for estimating effective population size in livestock assessed using data from the endangered black-coated Asturcón pony. *J. Anim. Sci.* **89,** 1251–1259.

Gracey, A. Y. (2007). Interpreting physiological responses to environmental change through gene expression profiling. *J. Exp. Biol.* **210,** 1584–1592.

Grado-Ahuir, J. A., Aad, P. Y., and Spicer, L. J. (2011). New insights into the pathogenesis of cystic follicles in cattle: Microarray analysis of gene expression in granulosa cells. *J. Anim. Sci.* **89,** 1769–1786.

Grant, W. S. (2006). Status and trends in genetic resources of capture fisheries. *In* "Workshop on Status and Trends in Aquatic Genetic Resources: A basis for international policy" (D. M. Bartley, B. J. Harvey, and R. S. V. Pullin, eds.), pp. 29–80. FAO, Rome.

Grattapaglia, D. (2008a). Perspectives on genome mapping and marker-assisted breeding of eucalypts. *South. Forests* **70,** 69–75.

Grattapaglia, D. (2008b). Genomics of *Eucalyptus*, a global tree for energy, paper, and wood. *In* "Genomics of Tropical Crop Plants" (P. H. Moore and R. Ming, eds.), Plant Genetics and Genomics: Crops and Models, Vol. 1, pp. 259–298. Springer, New York.

Grattapaglia, D., and Kirst, M. (2008). *Eucalyptus* applied genomics: From gene sequences to breeding tools. *New Phytol.* **179,** 911–929.

Grattapaglia, D., and Resende, M. D. V. (2011). Genomic selection in forest tree breeding. *Tree Genet. Genomes* **7,** 241–255.

Grattapaglia, D., Plomion, C., Kirst, M., and Sederoff, R. R. (2009). Genomics of growth traits in forest trees. *Curr. Opin. Plant Biol.* **12,** 148–156.

Griffiths, A. M., Machado-Schiaffino, G., Dillane, E., Coughlan, J., Horreo, J. L., Bowkett, A. E., Minting, P., Toms, S., Roche, W., Gargan, P., *et al.* (2010). Genetic stock identification of Atlantic salmon (*Salmo salar*) populations in the southern part of the European range. *BMC Genet.* **11,** 31.

Groenen, M. A. M., Wahlberg, P., Foglio, M., Cheng, H. H., Megens, H., Crooijmans, R. P. M. A., Besnier, F., Lathrop, M., Muir, W. M., Wong, G. K., *et al.* (2009). A high-density SNP-based linkage map of the chicken genome reveals sequence features correlated with recombination rate. *Genome Res.* **19,** 510–519.

Groenen, M. A. M., Amaral, A., Megens, H. J., Larson, G., Archibald, A. L., Muir, W. M., Malhi, R., Crooijmans, R. M. A., Ferretti, L., Perez-Encizo, M., *et al.* (2010). The Porcine HapMap Project: Genomewide assessment of nucleotide diversity, haplotype diversity and footprints of selection in the pig. W609. The International Plant & Animal Genome XVIII Conference. January 9-13, San Diego, California.

Groeneveld, E., Tinh, N. H., Kues, W., and Vien, N. T. (2008). A protocol for the cryoconservation of breeds by low-cost emergency cell banks—A pilot study. *Animal* **2,** 1–8.

Groeneveld, L. F., Lenstra, J. A., Eding, H., Toro, M. A., Scherf, B., Pilling, D., Negrini, R., Jianlin, H., Finlay, E. K., Groeneveld, E., *et al.* (2010). Genetic diversity in livestock breeds. *Anim. Genet.* **41**(Suppl. 1), 6–31.

Grönlund, A., Bhalerao, R. P., and Karlsson, J. (2009). Modular gene expression in Poplar: A multilayer network approach. *New Phytol.* **181,** 315–322.

Groover, A., Fontana, J., Dupper, G., Ma, C., Martienssen, R., Strauss, S., and Meilan, R. (2004). Gene and enhancer trap tagging of vascular expressed genes in poplar trees. *Plant Physiol.* **134,** 1742–1751.

Gross, B. L., and Olsen, K. M. (2010). Genetic perspectives on crop domestication. *Trends Plant Sci.* **15,** 529–537.

Grubman, M. J. (2005). Development of novel strategies to control foot-and-mouth disease: Marker vaccines and antivirals. *Biologicals* **33,** 227–234.

Guan, L., Yang, T., Li, N., Li, B., and Lu, H. (2010). Identification of superior clones by RAPD technology in *Xanthoceras sorbifolia* Bge. *For. Stud. China* **12,** 37–40.

Guedes, C. M., Gonçalves, D., Rodrigues, M. A. M., and Dias-Da-Silva, A. (2008). Effects of a *Saccharomyces cerevisiae* yeast on ruminal fermentation and fibre degradation of maize silages in cows. *Anim. Feed Sci. Technol.* **145,** 27–40.

Gueye, T., and Ndir, K. N. (2010). *In vitro* production of double haploid plants from two rice species (*Oryza sativa* L. and *Oryza glaberrima* Steudt.) for the rapid development of new breeding material. *Sci. Res. Essays* **5,** 709–713.

Guimarães, B. C. M., Arends, J. B. A., van der Ha, D., Van de Wiele, T., Boon, N., and Verstraete, W. (2010). Microbial services and their management: Recent progresses in soil bioremediation technology. *Appl. Soil Ecol.* **46,** 157–167.

Gupta, H. S., Agrawal, P. K., Mahajan, V. G., Bisht, S., Kumar, A., Verma, P., Srivastava, A., Saha, S., Babu, R., Pant, M. C., *et al.* (2009). Quality protein maize for nutritional security: Rapid development of short duration hybrids through molecular marker assisted breeding. *Curr. Sci.* **96,** 230–237.

Guy, C., Kopka, J., and Moritz, T. (2008). Plant metabolomics coming of age. *Physiol. Plant.* **132,** 113–116.

Gwo, J. C. (2000). Cryopreservation of sperm of some marine fishes. *In* "Cryopreservation in Aquatic Species" (T. R. Tiersch and P. M. Mazik, eds.), pp. 138–160. World Aquaculture Society, Baton Rouge.

Habte, M. (2000). Mycorrhizal fungi and plant nutrition. *In* "Plant Nutrient Management in Hawaii's Soils, Approaches for Tropical and Subtropical Agriculture" (J. A. Silva and R. Uchida, eds.), pp. 127–131. University of Hawaii, Manoa.

Haggman, H., Rusanen, M., and Jokipii, S. (2008). Cryopreservation of *in vitro* tissues of deciduous forest trees. *In* "Plant Cryopreservation—A Practical Guide" (B. B. M. Reed, ed.), pp. 365–386. Springer Science and Business Media, New York.

Hajjar, R., and Hodgkin, T. (2007). The use of wild relatives in crop improvement: A survey of developments over the last 20 years. *Euphytica* **156,** 1–13.

Halbert, N. D., Ward, T. J., Schnabe, R. D., Taylor, J. F., and Derr, J. N. (2005). Conservation genomics: Disequilibrium mapping of domestic cattle chromosomal segments in North American bison populations. *Mol. Ecol.* **14,** 2343–2362.

Hai, L., Wagner, C., and Friedt, W. (2007). Quantitative structure analysis of genetic diversity among spring bread wheats (*Triticum aestivum* L.) from different geographical regions. *Genetica* **130,** 213–225.

Hamaragodlu, F., Eroglu, A., Toner, M., and Sadler, K. C. (2005). Cryopreservation of starfish oocytes. *Cryobiology* **50,** 38–47.

Hamza, N. B. (2010). Cytoplasmic and nuclear DNA markers as powerful tools in populations' studies and in setting conservation strategies. *Afr. J. Biotechnol.* **9,** 4510–4515.

Hansen, P. J. (2006). Realizing the promise of IVF in cattle—An overview. *Theriogenology* **65,** 119–125.

Hao, C. Y., Zhang, X. Y., Wang, L. F., Dong, Y. S., Shang, X. W., and Jia, J. Z. (2006). Genetic diversity and core collection evaluations in common wheat germplasm from the Northwestern Spring Wheat Region in China. *Mol. Breed.* **17,** 69–77.

Hara, M., and Sekino, M. (2003). Efficient detection of parentage in a cultured Japanese flounder *Paralichthys olivaceus* using microsatellite DNA marker. *Aquaculture* **217,** 107–114.

Hara, K., Kon, Y., Sasazaki, S., Mukai, F., and Mannen, H. (2010). Development of novel SNP system for individual and pedigree control in a Japanese Black cattle population using whole-genome genotyping assay. *Anim. Sci. J.* **81,** 506–512.

Harding, K. (2004). Genetic integrity of cryopreserved plant cells: A review. *Cryo Letters* **25,** 3–22.

Harding, K. (2010). Plant and algal cryopreservation: Issues in genetic integrity, concepts in cryobionomics and current applications in cryobiology. *AsPac J. Mol. Biol. Biotechnol.* **18,** 151–154.

Harrison, E. J., Bush, M., Plett, J. M., McPhee, D. P., Vitez, R., O'Malley, B., Sharma, V., Bosnich, W., Séguin, A., MacKay, J., *et al.* (2007). Diverse developmental mutants revealed in an activation-tagged population of poplar. *Can. J. Bot.* **85,** 1071–1081.

Harvengt, L., Meier-Dinkel, A., Dumas, E., and Collin, E. (2004). Establishment of a cryopreserved genebank of European elms. *Can. J. For. Res.* **34,** 43–55.

Harvey, B. (1998). An overview of action before extinction. *In* "Action Before Extinction" (B. Harvey, C. Ross, D. Greer, and J. Carolsfeld, eds.), pp. 1–18. World Fisheries Trust, Vancouver.

Harvey, B. (2000). The application of cryopreservation on fish genetic conservation. *In* "Cryopreservation in Aquatic Species" (T. R. Tiersch and P. M. Mazik, eds.), Advances in World Aquaculture, Vol. 7, pp. 332–337. World Aquaculture Society, Baton Rouge.

Hashem, M. A. (2001). Problems and prospects of cyanobacterial biofertilizer for rice cultivation. *Aust. J. Plant Physiol.* **28,** 881–888.

Håstein, T., Gudding, R., and Evensen, O. (2005). Bacterial vaccines for fish—An update of the current situation worldwide. *Dev. Biol.* **121,** 55–74.

Hausman, J., Neys, O., Kevers, C., and Gasper, T. (1994). Effect of *in vitro* storage at 4°C on survival and proliferation of poplar shoots. *Plant Cell Tissue Organ Cult.* **38,** 65–67.

Hayes, B. J., Bowman, P. J., Chamberlain, A. C., and Goddard, M. E. (2009). Genomic selection in dairy cattle: Progress and challenges. *J. Dairy Sci.* **92,** 433–443.

Healy, T. M., Tymchuk, W. E., Osborne, E., and Schulte, P. M. (2010). Heat shock response of killifish (*Fundulus heteroclitus*): Candidate gene and heterologous microarray approaches. *Physiol. Genomics* **41,** 171–184.

Heaton, M. P., Harhay, G. P., Bennett, G. L., Stone, R. T., Grosse, W. M., Casas, E., Keele, J. W., Smith, T. P., Chitko-McKown, C. G., and Laegreid, W. W. (2002). Selection and use of SNP markers for animal identification and paternity analysis in U.S. beef cattle. *Mamm. Genome* **13,** 272–281.

Heffner, E. L., Sorrells, M. E., and Jannink, J. (2009). Genomic selection for crop improvement. *Crop Sci.* **49,** 1–12.

Heffner, E. L., Lorenz, A. J., Jannink, J. L., and Sorrells, M. E. (2010). Plant breeding with genomic selection: Potential gain per unit time and cost. *Crop Sci.* **50,** 1681–1690.

Heffner, E. L., Jannink, J. L., Iwata, H., Souza, E., and Sorrells, M. E. (2011). Genomic selection accuracy for grain quality traits in biparental wheat populations. *Crop Sci.* **51,** 2597–2606.

Heller, G., Adomas, A., Li, G., Osborne, J., van Zyl, L., Sederoff, R., Finlay, R. D., Stenlid, J., and Asiegbu, F. O. (2008). Transcriptional analysis of *Pinus sylvestris* roots challenged with the ectomycorrhizal fungus *Laccaria bicolor*. *BMC Plant Biol.* **8,** 19.

Henker, J., Muller, S., Laass, M. W., Schreiner, A., and Schulze, J. (2008). Probiotic *Escherichia coli* Nissle 1917 (EcN) for successful remission maintenance of ulcerative colitis in children and adolescents: An open-label pilot study. *Z. Gastroenterol.* **46,** 874–875.

Herlin, M., Delghandi, M., Wesmajervi, M., Taggart, J. B., McAndrew, B. J., and Penman, D. J. (2008). Analysis of the parental contribution to a group of fry from a single day of spawning from a commercial Atlantic cod (*Gadus morhua*) breeding tank. *Aquaculture* **274,** 218–224.

Hess, M., Sczyrba, A., Egan, R., Kim, T.-W., Chokhawala, H., Schroth, G., Luo, S., Clark, D. S., Chen, F., Zhang, T., *et al.* (2011). Metagenomic discovery of biomass-degrading genes and genomes from cow rumen. *Science* **331,** 463–467.

Heun, M., Schafer-Pregl, R., Klawan, D., Castagna, R., Accerbi, M., Borghi, B., and Salamini, F. (1997). Site of einkorn wheat domestication identified by DNA fingerprinting. *Science* **278,** 1312–1314.

Heyman, Y. (2010). From non surgical embryo transfer to somatic cloning in cattle: Technical challenges and hurdles to the use of reproductive biotechnologies. Proceedings of the 26th Annual Meeting A.E.T.E.—Kuopio, Finland, 10th-11th September. pp. 7–17.

Hiemstra, S. J., van der Lende, T., and Woelders, H. (2006). The potential of cryopreservation and reproductive technologies for animal genetic resources convervation strategies. *In* "The Role of Biotechnology in Exploring and Protecting Agricultural Genetic Resources" (J. Ruane and A. Sonnino, eds.), pp. 45–59. FAO, Rome.

Higgins, S. E., Higgins, J. P., Wolfenden, A. D., Henderson, S. N., Torres-Rodriguez, A., Tellez, G., and Hargis, B. (2008). Evaluation of a *Lactobacillus*-based probiotic culture for the reduction of *Salmonella Enteritidis* in neonatal broiler chicks. *Poult. Sci.* **87,** 27–31.

Hillel, J., Groenen, M. A., Tixier-Boichard, M., Korol, A. B., David, L., Kirzhner, V. M., Burke, T., Barre-Dirie, A., Crooijmans, R. P., Elo, K., *et al.* (2003). Biodiversity of 52 chicken populations assessed by microsatellite typing of DNA pools. *Genet. Sel. Evol.* **35,** 533–557.

Hirano, R., Jatoi, S. A., Kawase, M., Kikuchi, A., and Watanabe, K. N. (2009). Consequences of *ex situ* conservation on the genetic integrity of germplasm held at different gene banks: A case study of bread wheat collected in Pakistan. *Crop Sci.* **49,** 2160–2166.

Hiraoka, Y., Kuramoto, N., Okamura, M., Ohira, M., Taniguchi, T., and Fujisawa, Y. (2009). Clone identification and genetic relationship among candidates for superior trees in *Rhus succedanea* L. using ISSR, AFLP, and RAPD markers. *J. Jpn. For. Soc.* **91,** 246–252.

Hjelm, M., Bergh, O., Riaza, A., Nielsen, J., Melchiorsen, J., Jensen, S., Duncan, H., Ahrens, P., Birkbeck, H., and Gram, L. (2004). Selection and identification of autochthonous potential probiotic bacteria from turbot larvae (*Scophthalmus maximus*) rearing units. *Syst. Appl. Microbiol.* **27,** 360–371.

Hoban, S. M., McCleary, T. S., Schlarbaum, S. E., and Romero-Severson, J. (2009). Geographically extensive hybridization between the forest trees American butternut and Japanese walnut. *Biol. Lett.* **5,** 324–327.

Hocher, V., Alloisio, N., Auguy, F., Fournier, P., Doumas, P., Pujic, P., Gherbi, H., Queiroux, C., Da Silva, C., Wincker, P., *et al.* (2011). Transcriptomics of actinorhizal symbioses reveals homologs of the whole common symbiotic signaling cascade. *Plant Physiol.* **156,** 700–711.

Hoffmann, B., Beer, M., Reid, S. M., Mertens, P., Oura, C. A., van Rijn, P. A., Slomka, M. J., Banks, J., Brown, I. H., Alexander, D. J., *et al.* (2009). A review of RT-PCR technologies used in veterinary virology and disease control: Sensitive and specific diagnosis of five livestock diseases notifiable to the World Organisation for Animal Health. *Vet. Microbiol.* **139,** 1–23.

Holliday, J. A., Ralph, S. G., White, R., Bohlmann, J., and Aitken, S. N. (2008). Global monitoring of autumn gene expression within and among phenotypically divergent populations of Sitka spruce (*Picea sitchensis*). *New Phytol.* **178,** 103–122.

Hollingsworth, P. M., Dawson, I. K., Goodall-Copestake, W. P., Richardson, J. E., Weber, J. C., Sotelo Montes, C., and Pennington, R. T. (2005). Do farmers reduce genetic diversity when they domesticate tropical trees? A case study from Amazonia. *Mol. Ecol.* **142,** 497–501.

Holzapfel, W. H. (2002). Appropriate starter culture technologies for small-scale fermentation in developing countries. *Int. J. Food Microbiol.* **75,** 197–212.

Hook, S. E. (2010). Promise and progress in environmental genomics: A status report on the applications of gene expression-based microarray studies in ecologically relevant fish species. *J. Fish Biol.* **77,** 1999–2022.

Hoque, M. S., Broadhurst, L. M., and Thrall, P. H. (2011). Genetic characterization of root-nodule bacteria associated with *Acacia salicina* and *A. stenophylla* (Mimosaceae) across south-eastern Australia. *Int. J. Syst. Evol. Microbiol.* **61,** 299–309.

Horna, D., Debouck, D., Dumet, D., Hanson, J., Manyong, V. M., Payne, T., Sackville-Hamilton, R., Sanchez, I., Taba, S., and Upadhyaya, H. D. (2010). Costs effectiveness of germplasm collections in the CG system. Evaluating Cost-Effectiveness of Collection Management: *Ex-situ* Conservation of Plant Genetic Resources in the CG System. pp. 100–115. CGIAR.

Hornshoj, H., Bendixen, E., Conley, L. N., Andersen, P. K., Hedegaard, J., Panitz, F., and Bendixen, C. (2009). Transcriptomic and proteomic profiling of two porcine tissues using high-throughput technologies. *BMC Genomics* **10,** 30.

Houston, R. D., Haley, C. S., Hamilton, A., Guy, D. R., Tinch, A. E., Taggart, J. B., McAndrew, B. J., and Bishop, S. C. (2008). Major quantitative trait loci affect resistance to infectious pancreatic necrosis in Atlantic salmon (*Salmo salar*). *Genetics* **178,** 1109–1115.

Hu, Z., Fritz, E. R., and Reecy, J. M. (2007). AnimalQTLdb: A livestock QTL database tool set for positional QTL information mining and beyond. *Nucleic Acids Res.* **35,** D604–D609.

Hu, E., Yang, H., and Tiersch, T. R. (2011). High-throughput cryopreservation of spermatozoa of blue catfish (*Ictalurus furcatus*): Establishment of an approach for commercial-scale processing. *Cryobiology* **62,** 74–82.

Huang, X., Feng, Q., Qian, Q., Zhao, Q., Wang, L., Wang, A., Guan, J., Fan, D., Weng, Q., Huang, T., *et al.* (2009a). High-throughput genotyping by whole-genome resequencing. *Genome Res.* **19,** 1068–1076.

Huang, W. J., Ning, G. G., Liu, G. F., and Bao, M. Z. (2009b). Determination of genetic stability of long-term micropropagated plantlets of *Platanus acerifolia* using ISSR markers. *Biol. Plant.* **53,** 159–163.

Huang, M., Xie, F., Chen, L., Zhao, X., Jojee, L., and Madonna, D. (2010a). Comparative analysis of genetic diversity and structure in rice using ILP and SSR markers. *Rice Sci.* **17,** 257–268.

Huang, X., Wei, X., Sang, T., Zhao, Q., Feng, Q., Zhao, Y., Li, C., Zhu, C., Lu, T., Zhang, Z., *et al.* (2010b). Genome-wide association studies of 14 agronomic traits in rice landraces. *Nat. Genet.* **42,** 961–967.

Huang, W., Yandell, B. S., and Khatib, H. (2010c). Transcriptomic profiling of bovine IVF embryos revealed candidate genes and pathways involved in early embryonic development. *BMC Genomics* **11,** 23.

Hubalek, Z. (2003). Protectants used in the cryopreservation of microorganisms. *Cryobiology* **46,** 205–229.

Hubert, S., Higgins, B., Borza, T., and Bowman, S. (2010). Development of a SNP resource and a genetic linkage map for Atlantic cod (*Gadus morhua*). *BMC Genomics* **11,** 191.

Hulata, G. (2001). Genetic manipulations in aquaculture: A review of stock improvement by classical and modern technologies. *Genetica* **111,** 155–173.

Huo, X., Han, H., Zhang, J., and Yang, M. (2009). Genetic diversity of *Robinia pseudoacacia* populations in China detected by AFLP markers. *Front. Agric. China* **3,** 337–345.

IAEA (2008). Nuclear science for food security. IAEA press release. www.iaea.org/NewsCenter/PressReleases/2008/prn200820.html.

IAEA (2010). Nuclear technology review. http://www-pub.iaea.org/MTCD/publications/PDF/NTR2010_web.pdf.

Ibitoye, D. O., and Akin-Idowu, P. E. (2010). Marker-assisted-selection (MAS): A fast track to increase genetic gain in horticultural crop breeding. *Afr. J. Biotechnol.* **9,** 8889–8895.

International Chicken Genome Sequencing Consortium (2004). Sequence and comparative analysis of the chicken genome provide unique perspective on vertebrate evolution. *Nature* **432,** 695–716.

ISAG Conference (2008a). Cattle Molecular Markers and Parentage Testing Workshop. Amsterdam, The Netherlands.

ISAG Conference (2008b). Equine Genetics and Parentage Testing Standardization Workshop. Amsterdam, The Netherlands.

Jaccoud, D., Peng, K., Feinstein, D., and Kilian, A. (2001). Diversity arrays: A solid state technology for sequence information independent genotyping. *Nucleic Acids Res*. **29,** E25.

Jacques, R. J., Okeke, B. C., Bento, F. M., Peralba, M. C., and Camargo, F. A. (2009). Improved enrichment and isolation of Polycyclic Aromatic Hydrocarbons (PAH)-degrading micro-organisms in soil using anthracene as a model PAH. *Curr. Microbiol.* **58,** 628–634.

Jain, S. M. (2001). Tissue culture-derived variation in crop improvement. *Euphytica* **118,** 153–166.

Jalaja, N. C., Neelamathi, D., and Sreenivasan, T. V. (2008). Micropropagation for Quality Seed Production in Sugarcane in Asia and the Pacific. Food and Agriculture Organization of the United Nations, Rome. Asia–Pacific Consortium on Agricultural Biotechnology, New Delhi; Asia-Pacific Association of Agricultural Research Institutions, Bangkok.

Jalonen, R., Choo, K. Y., Hong, L. T., and Sim, H. C. (2009). Forest Genetic Resources Conservation and Management. Status in Seven South and Southeast Asian Countries. The Forest Research Institute Malaysia (FRIM), Bioversity International and APAFRI.

James, C. (2010). Global status of commercialized biotech/GM crops: 2010. http://www.isaaa.org/resources/publications/briefs/42/.

Jamnadass, R., Hanson, J., Poole, J., Hanotte, O., Simons, T. J., and Dawson, I. K. (2005). High differentiation among populations of the woody legume *Sesbania sesban* in sub-Saharan Africa: Implications for conservation and cultivation during germplasm introduction into agroforestry systems. *For. Ecol. Manage.* **210,** 225–238.

Jannink, J.-L., Lorenz, A. J., and Iwata, H. (2010). Genomic selection in plant breeding: From theory to practice. *Brief. Funct. Genomics* **9,** 166–177.

Jansen, J., and van Hintum, T. (2007). Genetic distance sampling: A novel sampling method for obtaining core collections using genetic distances with an application to cultivated lettuce. *Theor. Appl. Genet.* **114,** 421–428.

Jaramillo-Correa, J. P., Verdú, M., and González-Martínez, S. C. (2010). The contribution of recombination to heterozygosity differs among plant evolutionary lineages and life-forms. *BMC Evol. Biol.* **10,** 22.

Jarvis, A., Lane, A., and Hijmans, R. J. (2008). The effect of climate change on crop wild relatives. *Agric. Ecosyst. Environ.* **126,** 13–23.

Jasson, V., Jacxsens, L., Luning, P., Rajkovic, A., and Uyttendaele, M. (2010). Alternative microbial methods: An overview and selection criteria. *Food Microbiol.* **27,** 710–730.

Jauhar, P. P., Xu, S. S., and Baenziger, P. S. (2009). Haploidy in cultivated wheats: Induction and utility in basic and applied research. *Crop Sci.* **49,** 737–755.

Jenneckens, I., Müller-Belecke, A., Hörstgen-Schwark, G., and Meyer, J.-N. (1999). Proof of the successful development of Tilapia (*Oreochromis niloticus*) clones by DNA-fingerprinting. *Aquaculture* **173,** 377–388.

Jensen, S., Bergh, Ø., Enger, Ø., and Hjeltnes, B. (2002). Use of PCR–RFLP for genotyping 16S rRNA and characterizing bacteria cultured from halibut fry. *Can. J. Microbiol.* **48,** 379–386.

Jerry, D. R., Preston, N. P., Crocos, P. J., Keys, S., Meadows, J. R. S., and Li, Y. (2006). Application of DNA parentage analyses for determining relative growth rates of *Penaeus japonicus* families reared in commercial ponds. *Aquaculture* **254,** 171–181.

Jeukens, J., Renaut, S., St-Cyr, J., Nolte, A. W., and Bernatchez, L. (2010). The transcriptomics of sympatric dwarf and normal lake whitefish (*Coregonus clupeaformis* spp., Salmonidae) divergence as revealed by next-generation sequencing. *Mol. Ecol.* **19,** 5389–5403.

Jimenez, N., Vinas, M., Sabate, J., Diez, S., Bayona, J. M., Solanas, A. M., and Albaiges, J. (2006). The Prestige oil spill. 2. Enhanced biodegradation of a heavy fuel oil under field conditions by the use of an oleophilic fertilizer. *Environ. Sci. Technol.* **40,** 2578–2585.

Johansen, S. D., Karlsen, B. O., Furmanek, T., Andreassen, M., Jørgensen, T. E., Bizuayehu, T. T., Breines, R., Emblem, A., Kettunen, P., Luukko, K., *et al.* (2011a). RNA deep sequencing of the Atlantic cod transcriptome. *Comp. Biochem. Physiol. Part D Genomics Proteomics* **6,** 18–22.

Johansen, L.-H., Jensen, I., Mikkelsen, H., Bjørn, P.-A., Jansen, P. A., and Bergh, Ø. (2011b). Disease interaction and pathogens exchange between wild and farmed fish populations with special reference to Norway. *Aquaculture* **315,** 167–186.

John, U. P., and Spangenberg, G. C. (2005). Xenogenomics: Genomic bioprospecting in indigenous and exotic plants through EST discovery, cDNA microarray-based expression profiling and functional genomics. *Comp. Funct. Genomics* **6,** 230–235.

Johnston, J. W., Benson, E. E., and Harding, K. (2009). Cryopreservation induces temporal DNA methylation epigenetic changes and differential transcriptional activity in Ribes germplasm. *Plant Physiol. Biochem.* **47,** 123–131.

Jones, M. E., Shepherd, M., Henry, R. J., and Delves, A. (2006). Chloroplast DNA variation and population structure in the widespread forest tree, *Eucalyptus grandis*. *Conserv. Genet.* **7,** 691–703.

Jose, F. C., Mohammed, M. M. S., Thomas, G., Varghese, G., Selvaraj, N., and Dorai, M. (2009). Genetic diversity and conservation of common bean (*Phaseolus vulgaris* L., Fabaceae) landraces in Nilgiris. *Curr. Sci.* **97,** 227–235.

Jouany, J. P. (2006). Optimizing rumen functions in the close-up transition period and early lactation to drive dry matter intake and energy balance in cows. *Anim. Reprod. Sci.* **96,** 250–264.

Juhasz, A. G., Venczel, G., Sagi, Z., Gajdos, L., Zatyko, L., Kristof, Z., and Vagi, P. (2006). Production of doubled haploid breeding lines in case of paprika, spice paprika, eggplant, cucumber, zucchini and onion. *Acta Hortic.* **725,** 845–853.

Jung, J. Y., Seo, E., Jang, K., Kim, T.-J., Yoon, H. S., and Han, N. S. (2011). Monitoring of microbial changes in salted cabbage (Jeolimbaechu) during recycled brining operation. *Food Sci. Biotechnol.* **20,** 223–227.

Kabaluk, J. T., Svircev, A. M., Goettel, M. S., and Woo, S. G. (2010). *The Use and Regulation of Microbial Pesticides in Representative Jurisdictions Worldwide*. IOBC Global. http://www.iobc-global.org/downlaod/Microbial_Regulation_Book_Kabaluk_et_%20al_2010.pdf.

Kaity, A., Ashmore, S. E., Drew, R. A., and Dulloo, M. E. (2008). Assessment of genetic and epigenetic changes following cryopreservation in papaya. *Plant Cell Rep.* **27,** 1529–1539.

Kamm, U., Rotach, P., Gugerli, F., Siroky, M., Edwards, P., and Holderegger, R. (2009). Frequent long-distance gene flow in a rare temperate forest tree (*Sorbus domestica*) at the landscape scale. *Heredity* **103,** 476–482.

Kang, K. H., Zhang, Z., Bao, Z., and Shao, M. (2009). Cryopreservation of veliger larvae of trumpet shell, *Charonia sauliae*: An essential preparation to artificial propagation. *J. Ocean Univ. China* **8,** 265–269.

Karlsson, S., Moen, T., Lien, S., Glover, K. A., and Hindar, K. (2011). Generic genetic differences between farmed and wild Atlantic salmon identified from a 7K SNP-chip. *Mol. Ecol. Resour.* **11** (Suppl. 1), 247–253.

Kassahn, K. S. (2008). Microarrays for comparative and ecological genomics: Beyond single species applications of array technologies. *J. Fish Biol.* **72,** 2407–2434.

Kav, N. N. V., Srivastava, S., Yajima, W., and Sharma, N. (2007). Application of proteomics to investigate plant-microbe interactions. *Curr. Proteomics* **4,** 28–43.

Kawar, P. G., Devarumath, R. M., and Nerkar, Y. (2009). Use of RAPD markers for assessment of genetic diversity in sugarcane cultivars. *Indian J. Biotechnol.* **8,** 67–71.

Keller, E. R. J., Senula, A., Leunufna, S., and Grübe, M. (2006). Slow growth storage and cryopreservation—Tools to facilitate germplasm maintenance of vegetatively propagated crops in living plant collections. *Int. J. Refrig.* **29,** 411–417.

Kerrigan, L. (2007). Cryopreservation of bacteria. *PMF Newsl.* **13,** 2–6.

Khan, A. A., Jilani, G., Akhtar, M. S., Naqvi, S. M. S., and Rasheed, M. (2009a). Phosphorus solubilizing bacteria: Occurrence, mechanisms and their role in crop production. *J. Agric. Biol. Sci.* **1,** 48–58.

Khan, M. S., Zaidi, A., and Wani, P. A. (2009b). Role of phosphate solubilizing micro-organisms in sustainable agriculture—A review. *In* "Sustainable Agriculture 2009" (E. Lichtfouse, M. Navarrete, P. Debaeke, S. Véronique, and C. Alberola, eds.), pp. 551–570. Springer, Dordrecht, Heidelberg, London, New York.

Khan, M. S., Zaidi, A., Ahemad, M., Oves, M., and Wani, P. A. (2010). Plant growth promotion by phosphate solubilizing fungi—Current perspective. *Arch. Agronomy Soil Sci.* **56,** 73–98.

Kijima, Y., Otsuka, K., and Sserunkuuma, D. (2008). Assessing the impact of NERICA on income and poverty in central and western Uganda. *Agric. Econ.* **38,** 327–337.

Kijima, Y., Otsuka, K., and Sserunkuuma, D. (2011). An inquiry into constraints on a green revolution in sub-Saharan Africa: The case of NERICA rice in Uganda. *World Dev.* **39,** 77–86.

Kim, S. W. (2010). Bio-fermentation technology to improve efficiency of swine nutrition. *Asian-Aust. J. Anim. Sci.* **23,** 825–832.

Kim, J. S., Lee, G. G., Park, J. S., Jung, Y. H., Kwak, H. S., Kim, S. B., Nam, Y. S., and Kwon, S. T. (2007). A novel multiplex PCR assay for rapid and simultaneous detection of five pathogenic bacteria: *Escherichia coli* O157: H7, *Salmonella, Staphylococcus aureus, Listeria monocytogenes*, and *Vibrio parahaemolyticus*. *J. Food Prot.* **70,** 1656–1662.

Kim, H. H., Shin, D. J., No, N. Y., Yoon, M. K., Choi, H. S., Lee, J. S., and Engelmann, F. (2009). Cryopreservation of garlic germplasm collections using the droplet–vitrification technique. *Abst. 1st international symposium on cryopreservation in horticultural species, Leuven, Belgium, 39, 5–8 April, 2009.*

Kim, M. J., An, H. S., and Choi, K. H. (2010a). Genetic characteristics of Pacific cod populations in Korea based on microsatellite markers. *Fish. Sci.* **76,** 595–603.

Kim, S., Eo, H. S., Koo, H., Choi, J. K., and Kim, W. (2010b). DNA barcode-based molecular identification system for fish species. *Mol. Cells* **30,** 507–512.

Kindt, R. Muchugi, A., Hansen, O. K., Kipruto, H., Poole, J., Dawson, I., and Jamnadass, R. (2009). *Molecular Markers for Tropical Trees: Statistical Analysis of Dominant Data. ICRAF Technical Manual No. 13.* The World Agroforestry Centre, Nairobi.

King, T., Lucchini, S., Hinton, J. C. D., and Gobius, K. (2010). Transcriptomic analysis of *Escherichia coli* O157:H7 and K-12 cultures exposed to inorganic and organic acids in stationary phase reveals acidulant- and strain-specific acid tolerance responses. *Appl. Environ. Microbiol.* **76,** 6514–6528.

Kirst, M., Myburg, A. A., Kirst, M. E., Scott, J., and Sederoff, R. R. (2004). Coordinated genetic regulation of growth and lignin content revealed by QTL analysis of cDNA microarray data in an interspecific cross of *Eucalyptus*. *Plant Physiol.* **135,** 2368–2378.

Kisa, M., Sanson, A., Thioulouse, J., Assigbetse, K., Sylla, S., Spichiger, R., Dieng, L., Berthelin, J., Prin, Y., Galiana, A., *et al.* (2007). Arbuscular mycorrhizal symbiosis can counterbalance the negative influence of the exotic tree species *Eucalyptus camaldulensis* on the structure and functioning of soil microbial communities in a sahelian soil. *FEMS Microbiol. Ecol.* **62,** 32–44.

Kitada, S., Shishidou, H., Sugaya, T., Kitakado, T., Hamasaki, K., and Kishino, H. (2009). Genetic effects of long-term stock enhancement programs. *Aquaculture* **290,** 69–79.

Kleerebezem, M., and de Vos, W. M. (2011). Lactic acid bacteria: Life after genomics. *Microb. Biotechnol.* **4,** 318–322.

Kloosterman, B., Oortwijn, M., uitdewilligen, J., America, T., de Vos, R., Visser, R. G., and Bachem, C. W. (2010). From QTL to candidate gene: Genetical genomics of simple and complex traits in potato using a pooling strategy. *BMC Genomics* **11,** 158–173.

Kochzius, M., Nolte, M., Weber, H., Silkenbeumer, N., Hjörleifsdottir, S., Hreggvidsson, G. O., Marteinsson, V., Kappel, K., Planes, S., Tinti, F., *et al.* (2008). DNA microarrays for identifying fishes. *Mar. Biotechnol.* **10,** 207–217.

Kochzius, M., Seidel, C., Antoniou, A., Botla, S. K., Campo, D., Cariani, A., Vazquez, E. G., Hauschild, J., Hervet, C., Hjörleifsdottir, S., *et al.* (2010). Identifying fishes through DNA barcodes and microarrays. *PLoS One* **5,** e12620.

Komen, H., and Thorgaard, G. (2007). Androgenesis, gynogenesis and the production of clones in fishes: A review. *Aquaculture* **269,** 150–173.

Kon, T., Yoshino, T., Mukai, T., and Nishida, M. (2007). DNA sequences identify numerous cryptic species of the vertebrate: A lesson from the gobioid fish *Schindleria. Mol. Phylogenet. Evol.* **44,** 53–62.

Konstantinov, S. R., Smidt, H., Akkermans, A. D. L., Casini, L., Trevisi, P., Mazzoni, M., De Filippi, S., Bosi, P., and de Vos, W. (2008). Feeding of *Lactobacillus sobrius* reduces *Escherichia coli* F4 levels in the gut and promotes growth of infected piglets. *FEMS Microbiol. Ecol.* **66,** 599–607.

Kostic, T., Francois, P., Bodrossy, L., and Schrenzel, J. (2008). Oligonucleotide and DNA microarrays: Versatile tools for rapid bacterial diagnostics. *In* "Principles of Bacterial Detection: Biosensors, Recognition Receptors and Microsystems" (M. Zourob, S. Elwary, and A. Turner, eds.), pp. 629–657. Springer, New York.

Krasnov, A., Timmerhaus, G., Afanasyev, S., and Jørgensen, S. M. (2011). Development and assessment of oligonucleotide microarrays for Atlantic salmon (*Salmo salar* L.). *Comp. Biochem. Physiol. Part D Genomics Proteomics* **6,** 31–38.

Kreuzwieser, J., Hauberg, J., Howell, K. A., Carroll, A., Rennenberg, H., Millar, A. H., and Whelan, J. (2009). Differential response of gray poplar leaves and roots underpins stress adaptation during hypoxia. *Plant Physiol.* **149,** 461–473.

Krivanek, A. F., De Grote, H., Gunaratna, N. S., Diallo, A. O., and Friesen, D. (2007). Breeding and disseminating quality protein maize (QPM) for Africa. *Afr. J. Biotechnol.* **6,** 312–324.

Krutovsky, V., Troggio, M., Brown, G. R., Jermstad, K. D., and Neale, D. B. (2004). Comparative mapping in the Pinaceae. *Genetics* **168,** 447–461.

Kues, W. A., and Niemann, H. (2004). The contribution of farm animals to human health. *Trends Biotechnol.* **22,** 286–294.

Kugonza, D. R., Jianlin, H., Nabasirye, M., Mpairwea, D., Kiwuwa, G. H., Okeyo, A. M., and Hanotte, O. (2011). Genetic diversity and differentiation of Ankole cattle populations in Uganda inferred from microsatellite data. *Livest. Sci.* **135,** 140–147.

Kuhl, H., Tine, M., Hecht, J., Knaust, F., and Reinhardt, R. (2011). Analysis of single nucleotide polymorphisms in three chromosomes of European sea bass *Dicentrarchus labrax. Comp. Biochem. Physiol. Part D Genomics Proteomics* **6,** 70–75.

Külheim, C., Yeoh, S. H., Maintz, J., Foley, W. J., and Moran, G. F. (2009). Comparative SNP diversity among four *Eucalyptus* species for genes from secondary metabolite biosynthetic pathways. *BMC Genomics* **10,** 452.

Kumar, V. V.S, and Biji, C.P (2009). Biocontrol of the teak defoliator Hyblaea puera Prospects and constraints. *ENVIS Forestry Bulletin* 9. http://www.frienvis.nic.in/Bulletinwork/EFB-2009-I.htm.

Kumar, V. J., Achuthan, C., Manju, N. J., Philip, R., and Singh, I. S. B. (2009). Stringed bed suspended bioreactors (SBSBR) for *in situ* nitrification in penaeid and non-penaeid hatchery systems. *Aquacult. Int.* **17,** 479–489.

Kumar, G. R., Sakthivel, K., Sundaram, R. M., Neeraja, C. N., Balachandran, S. M., Rani, N. S., Viraktamath, B. C., and Madhav, M. S. (2010). Allele mining in crops: Prospects and potentials. *Biotechnol. Adv.* **28**(4), 451–461.

Kumar, K., Desai, V., Cheng, L., Khitrov, M., Grover, D., Satya, R. V., Yu, C., Zavaljevski, N., and Reifman, J. (2011). AGeS: A software system for microbial genome sequence annotation. *PLoS One* **6,** e17469.

Kumari, N., Srivastava, A. K., Bhargava, P., and Rai, L. C. (2009). Molecular approaches towards assessment of cyanobacterial biodiversity. *Afr. J. Biotechnol.* **8,** 4284–4298.

Kurath, G. (2008). Biotechnology and DNA vaccines for aquatic animals. *Rev. Sci. Tech.* **27,** 175–196.

Kurokura, H., and Oo, K. M. (2008). Evaluation of fertilizing capacity of cryopreserved rainbow trout sperm. *Fish. Sci.* **74,** 621–626.

Kuzyk, M. A., Burian, J., Machander, D., Dolhaine, D., Cameron, S., Thornton, J. C., and Kay, W. W. (2001). An efficacious recombinant subunit vaccine against the salmonid rickettsial pathogen *Piscirickettsia salmonis*. *Vaccine* **19,** 2337–2344.

Kwantong, S., and Bart, A. N. (2009). Fertilization efficiency of cryopreserved sperm from striped catfish, *Pangasius hypophthalmus* (Sauvage). *Aquacult. Res.* **40,** 292–297.

Lacerda, S., Batlouni, S., Assis, L., Resende, F., Campos-Silva, S., Campos-Silva, R., Segatelli, T., and Franca, L. (2008). Germ cell transplantation in tilapia (*Oreochromis niloticus*). *Cybium* **32,** 115–118.

Lagoda, P. J. L. (2009). Networking and fostering of cooperation in plant mutation genetics and breeding: Role of the Joint FAO/IAEA Division. *In* "Induced Plant Mutations in the Genomics Era, Proceedings of an International Joint FAO/IAEA Symposium, 2008" (Q. Y. Shu, ed.), pp. 27–30. FAO, Rome.

Lanaud, C., Fouet, O., Clement, D., Boccara, M., Risterucci, A. M., Surujdeo-Maharaj, S., Legavre, T., and Argout, X. (2009). A meta-QTL analysis of disease resistance traits of *Theobroma cacao* L. *Mol. Breed.* **24,** 361–374.

Lane, E. A., Austin, E. J., and Crowe, M. A. (2008). Oestrous synchronisation in cattle—Current options following the EU regulations restricting use of oestrogenic compounds in food-producing animals: A review. *Anim. Reprod. Sci.* **109,** 1–16.

Langridge, P., and Fleury, D. (2011). Making the most of 'omics' for crop breeding. *Trends Biotechnol.* **29,** 33–40.

Lanteri, S., and Barcaccia, S. (2006). Molecular marker based analysis for crop germplasm conservation. *In* "The Role of Biotechnology in Exploring and Protecting Agricultural Genetic Resources" (J. Ruane and A. Sonnino, eds.), pp. 105–120. FAO, Rome.

Lara-Flores, M., Olvera-Novoa, M. A., Guzman-Mendez, B. E., and Lopez-Madrid, W. (2003). Use of the bacteria *Streptococcus faecium* and *Lactobacillus acidophilus*, and the yeast *Saccharomyces cerevisiae* as growth promoters in Nile tilapia (*Oreochromis niloticus*). *Aquaculture* **216,** 193–201.

Larsen, P. F., Nielsen, E. E., Williams, T. D., Hemmer-Hansen, J., Chipman, J. K., Kruhøffer, M., Grønkjær, P., George, S. G., Dyrskjøt, L., and Loeschcke, V. (2007). Adaptive differences in gene expression in European flounder (*Platichthys flesus*). *Mol. Ecol.* **16,** 4674–4683.

Larsen, P. F., Schulte, P. M., and Nielsen, E. E. (2011). Gene expression analysis for the identification of selection and local adaptation in fishes. *J. Fish Biol.* **78,** 1–22.

Leary, D. K., and Walton, D. W. H. (2010). Science for profit. What are the ethical implications of bioprospecting in the Arctic and Antarctica? *Ethics Sci. Environ. Polit.* **10,** 1–4.

Le Calvez, T., Burgaud, G., Mahe, S., Barbier, G., and Vandenkoornhuyse, P. (2009). Fungal diversity in deep-sea hydrothermal ecosystems. *Appl. Environ. Microbiol.* **75,** 6415–6421.

Le Cunff, L., Fournier-Level, A., Laucou, V., Vezzulli, S., Lacombe, T., Adam-Blondon, A. F., Boursiquot, J. M., and This, P. (2008). Construction of nested genetic core collections to optimize the exploitation of natural diversity in *Vitis vinifera* L. subsp. sativa. *BMC Plant Biol.* **8,** 31–42.

Lee, S., and Moorman, G. (2008). Identification and characterization of simple sequence repeat markers for *Pythium aphanidermatum*, *P. cryptoirregulare*, and *P. irregulare* and the potential use in *Pythium* population genetics. *Curr. Genet.* **53,** 81–93.

Lengkeek, A. G., Mwangi, A. M., Agufa, C. A. C., Ahenda, J. O., and Dawson, I. K. (2006). Comparing genetic diversity in agroforestry systems with natural forest: A case study of the important timber tree *Vitex fischeri* in central Kenya. *Agrofor. Syst.* **67,** 293–300.

Leong, J. S., Jantzen, S. G., von Schalburg, K. R., Cooper, G. A., Messmer, A. M., Liao, N. Y., Munro, S., Moore, R., Holt, R. A., Jones, S. J., *et al.* (2010). *Salmo salar* and *Esox lucius* full-length cDNA sequences reveal changes in evolutionary pressures on a post-tetraploidization genome. *BMC Genomics* **11,** 279.

Leroy, G., Callède, L., Verrier, E., Mériaux, J.-C., Ricard, A., Danchin-Burge, C., and Rognon, X. (2009). Genetic diversity of a large set of horse breeds raised in France assessed by microsatellite polymorphism. *Genet. Sel. Evol.* **41,** 5–16.

Lexer, C., Heinze, B., Alia, R., and Rieseberg, L. H. (2004). Hybrid zones as a tool for identifying adaptive genetic variation in outbreeding forest trees: Lessons from wild annual sunflowers (*Helianthus* spp.). *For. Ecol. Manage.* **197,** 49–64.

Lexer, C., Fay, M. F., Joseph, J. A., Nica, M. S., and Heinze, B. (2005). Barrier to gene flow between two ecologically divergent *Populus* species, *P. alba* (white poplar) and *P. tremula* (European aspen): The role of ecology and life history in gene introgression. *Mol. Ecol.* **14,** 1045–1057.

Li, Z. Z. (2007). *Beauveria bassiana* for pine caterpillar management in the People's Republic of China. *In* "Biological Control: A Global Perspective" (C. Vincent, M. S. Goettel, and G. Lazarovits, eds.), pp. 300–310. CABI Publishing, Oxfordshire.

Li, B. L., Zhang, Y. L., Wang, H., Song, C. H., and Liu, Y. (2009a). Pollen cryo-bank establishment and application of traditional Chinese flowers. *Abst. CRYO '09, Annual meeting of the Society for Cryobiology, Tsukuba, Japan, 108, 21–26 July, 2009.* .

Li, X. J., Lin, X., Li, P. J., Liu, W., Wang, L., Ma, F., and Chukwuka, K. S. (2009b). Biodegradation of the low concentration of polycyclic aromatic hydrocarbons in soil by microbial consortium during incubation. *J. Hazard. Mater.* **172,** 601–605.

Li, L. H., Qiu, X. H., Li, X. H., Wang, S. P., Zhang, Q. F., and Lian, X. M. (2010a). Transcriptomic analysis of rice responses to low phosphorus stress. *Chin. Sci. Bull.* **55,** 251–258.

Li, S., Pozhitkov, A., Ryan, R. A., Manning, C. S., Brown-Peterson, N., and Brouwer, M. (2010b). Constructing a fish metabolic network model. *Genome Biol.* **11**(11), R115.

Li, J., Boroevich, K. A., Koop, B. F., and Davidson, W. S. (2011). Comparative genomics identifies candidate genes for Infectious Salmon Anemia (ISA) resistance in Atlantic salmon (*Salmo salar*). *Mar. Biotechnol.* **13,** 232–241.

Ling, Y. H., Ma, Y. H., Guan, W. J., Cheng, Y. J., Wang, Y. P., Han, J. L., Mang, L., Zhao, Q. J., He, X. H., Pu, Y. B., *et al.* (2010). Evaluation of the genetic diversity and population structure of Chinese indigenous horse breeds using 27 microsatellite markers. *Anim. Genet.* **42,** 56–65.

Liu, Z. (2007). Fish genomics and analytical genetic technologies, with examples of their potential applications in management of fish genetic resources. *In* "Workshop on Status and Trends in Aquatic Genetic Resources: A Basis for International Policy" (D. M. Bartley, B. J. Harvey, and R. S. V. Pullin, eds.), pp. 145–179. FAO, Rome.

Liu, Z. J., and Cordes, J. F. (2004). DNA marker technologies and their applications in aquaculture genetics. *Aquaculture* **238,** 1–37.

Liu, J., Xu, X., and Deng, X. (2005). Intergeneric somatic hybridization and its application to crop genetic improvement. *Plant Cell Tissue Organ Cult.* **82,** 19–44.

Liu, Y.-G., Liu, L.-X., Wang, L., and Gao, A.-Y. (2008). Determination of genetic stability in surviving apple shoots following cryopreservation by vitrification. *Cryo Letters* **29,** 7–14.

Liu, W. X., Luo, Y. M., Teng, Y., Li, Z. G., and Ma, L. Q. (2010). Bioremediation of oily sludge-contaminated soil by stimulating indigenous microbes. *Environ. Geochem. Health* **32,** 23–29.

Liu, S., Zhou, Z., Lu, J., Sun, F., Wang, S., Liu, H., Jiang, Y., Kucuktas, H., Kaltenboeck, L., Peatman, E., *et al.* (2011a). Generation of genome-scale gene-associated SNPs in catfish for the construction of a high-density SNP array. *BMC Genomics* **12,** 53.

Liu, S., Han, Y., and Zhou, Z.-H. (2011b). Lactic acid bacteria in traditional fermented Chinese foods. *Food Res. Int.* **44,** 643–651.

Loftus, R. T., Ertugrul, O., Harba, A. H., El-Barody, M. A., MacHugh, D. E., Park, S. D., and Bradley, D. G. (1999). A microsatellite survey of cattle from a centre of origin: The Near East. *Mol. Ecol.* **8,** 2015–2022.

Long, J. A. (2008). Reproductive biotechnology and gene mapping: Tools for conserving rare breeds of livestock. *Reprod. Domest. Anim.* **43**(Suppl. 2), 83–88.

Lopes, T., Pinto, G., Loureiro, J., Costa, A., and Santos, C. (2006). Determination of genetic stability in long-term somatic embryogenic cultures and derived plantlets of cork oak using microsatellite markers. *Tree Physiol.* **26,** 1145–1152.

Loukovitis, D., Sarropoulou, E., Tsigenopoulos, C. S., Batargias, C., Magoulas, A., Apostolidis, A. P., Chatziplis, D., and Kotoulas, G. (2011). Quantitative trait loci involved in sex determination and body growth in the gilthead sea bream (*Sparus aurata* L.) through targeted genome scan. *PLoS One* **6,** 16599.

Lu, J., Peatman, E., Yang, Q., Wang, S., Hu, Z., Reecy, J., Kucuktas, H., and Liu, Z. (2011). The catfish genome database cBARBEL: An informatic platform for genome biology of ictalurid catfish. *Nucleic Acids Res.* **39,** D815–D821.

Lucy, M. C., McDougall, S., and Nation, D. P. (2004). The use of hormonal treatments to improve the reproductive performance of lactating dairy cows in feedlot or pasture-based management systems. *Anim. Reprod. Sci.* **82–83,** 495–512.

Ludwig, A. (2008). Identification of *Acipenseriformes* species in trade. *J. Appl. Ichthyol.* **24**(Suppl. 1), 2–19.

MacAvoy, E. S., Wood, A. R., and Gardner, J. P. A. (2008). Development and evaluation of microsatellite markers for identification of individual Greenshell^{TM} mussels (*Perna canaliculus*) in a selective breeding programme. *Aquaculture* **274,** 41–48.

Maccaferri, M., Sanguineti, M. C., Donini, P., and Tuberosa, R. (2003). Microsatellite analysis reveals a progressive widening of the genetic basis in the elite durum wheat germplasm. *Theor. Appl. Genet.* **107,** 783–797.

Mace, E. S., Rami, J. F., Bouchet, S., Klein, P. E., Klein, R. R., Kilian, A., Wenzl, P., Xia, L., Halloran, K., and Jordan, D. R. (2009). A consensus genetic map of sorghum that integrates multiple component maps and high-throughput Diversity Array Technology (DArT) markers. *BMC Plant Biol.* **9,** 13.

MacKenzie, A. A. (2005). *Applications of genetic engineering for livestock and biotechnology products, Technical Item II, 73rd General Session, Paris, International Committee, OIE.* ftp://ftp.fao.org/codex/ccfbt5/bt0503ae.pdf.

MacKinnon, K. M., Burton, J. L., Zajac, A. M., and Notter, D. R. (2009). Microarray analysis reveals difference in gene expression profiles of hair and wool sheep infected with Haemonchus contortus. *Vet. Immunol. Immunopathol.* **130,** 210–220.

Madoroba, E., Steenkamp, E. T., Theron, J., Scheirlinck, I., Cloete, T. E., and Huys, G. (2011). Diversity and dynamics of bacterial populations during spontaneous sorghum fermentations used to produce ting, a South African food. *Syst. Appl. Microbiol.* **34,** 227–234.

Magyary, I., Urbanyi, B., Horvath, A., and Dinnyes, A. (2000). Cryopreservation of gametes and embryo of cyprinid fishes. *In* "Cryopreservation in Aquatic Species" (T. R. Tiersch and P. M. Mazik, eds.), pp. 199–210. World Aquaculture Society, Baton Rouge.

Majhi, S. K., Hattori, R. S., Yokota, M., Watanabe, S., and Strüssmann, C. A. (2009). Germ cell transplantation using sexually competent fish: An approach for rapid propagation of endangered and valuable germlines. *PLoS One* **4,** e6132.

Malorny, B., Lofstrom, C., Wagner, M., Kramer, N., and Hoorfar, J. (2008). Enumeration of *Salmonella* bacteria in food and feed samples by real-time PCR for quantitative microbial risk assessment. *Appl. Environ. Microbiol.* **74,** 1299–1304.

Maluszynski, M., Kasha, K. J., Forster, B. P., and Szarejko, I. (2003). Doubled Haploid Production in Crop Plants: A Manual. Kluwer Academic Publishers, The Netherlands.

Manaa, A., Ben Ahmed, H., Valot, B., Bouchet, J. P., Aschi-Smiti, S., Causse, M., and Faurobert, M. (2011). Salt and genotype impact on plant physiology and root proteome variations in tomato. *J. Exp. Bot.* **62,** 2797–2813.

Mandal, B. B. (2000). Cryopreservation research in India: Current status and future perspectives. *In* "Cryopreservation of Tropical Plant Germplasm—Current Research Progress and Applications" (F. Engelmann and H. Takagi, eds.), pp. 282–286. JIRCAS, Tsukuba.

Mandal, P. K., Biswas, A. K., Choi, K., and Pal, U. K. (2011). Methods for rapid detection of foodborne pathogens: An overview. *Am. J. Food Technol.* **6,** 87–102.

Manju, N. J., Deepesh, V., Achuthan, C., Rosamma, P., and Singh, I. S. B. (2009). Immobilization of nitrifying bacterial consortia on wood particles for bioaugmenting nitrification in shrimp culture systems. *Aquaculture* **294,** 65–75.

Manju, R. A., Haniffa, M. A., Singh, S. V. A., Ramakrishnan, C. M., Dhanaraj, M., Innocent, B. X., Seetharaman, S., and Arockiaraj, A. J. (2011). Effect of dietary administration of Efinol® FG on growth and enzymatic activities of *Channa striatus* (Bloch, 1793). *J. Anim. Vet. Adv.* **10,** 796–801.

Manning, A. J., Burton, M. P. M., and Crim, L. W. (2004). Reproductive evaluation of triploid yellowtail flounder, *Limanda ferruginea* (Storer). *Aquaculture* **242,** 625–640.

Mantegazza, R., Biloni, M., Grassi, F., Basso, B., Lu, B. R., Cai, X. X., Sala, F., and Spada, A. (2008). Temporal trends of variation in Italian rice germplasm over the past two centuries revealed by AFLP and SSR markers. *Crop Sci.* **48,** 1832–1840.

Marco, D. (2010). Metagenomics: Theory, Methods and Applications. Caister Academic Press, Norfolk.

Marita, J. M., Rodriguez, J. M., and Nienhuis, J. (2000). Development of an algorithm identifying maximally diverse core collections. *Genet. Resour. Crop Evol.* **47,** 515–526.

Márquez, G. C., Siegel, P. B., and Lewis, R. M. (2010). Genetic diversity and population structure in lines of chickens divergently selected for high and low 8-week body weight. *Poult. Sci.* **89,** 2580–2588.

Marshall, K., Quiros-Campos, C., van der Werf, J. H. J., and Kinghorn, B. (2011). Marker-based selection within smallholder production systems in developing countries. *Livest. Sci.* **136,** 45–54.

Martin, M. T., Pedranzani, H. E., and de Grado, R. S. (2007). Behavior and preservation of an *in vitro* collection of European aspen in Spain. *Biocell* **31,** 41–49.

Mashkina, O. S., Tabatskaya, T. M., Gorobets, A. I., and Shestibratov, K. A. (2010). Method of clonal micropropagation of different willow species and hybrids. *Appl. Biochem. Microbiol.* **46,** 769–775.

Massault, C., Franch, R., Haley, C., De Koning, D. J., Bovenhuis, H., Pellizzari, C., Patarnello, T., and Bargelloni, L. (2011). Quantitative trait loci for resistance to fish pasteurellosis in gilthead sea bream (*Sparus aurata*). *Anim. Genet.* **42,** 191–203.

Mathesius, U. (2009). Comparative proteomic studies of root-microbe interactions. *J. Proteomics* **72,** 353–366.

Mathur, J., Mukunthakumar, S., Gupta, S. N., and Mathur, S. N. (1991). Growth and morphogenesis of plant tissue cultures under mineral-oil. *Plant Sci.* **74,** 249–254.

Matsumoto, C. K., and Hilsdorf, A. W. S. (2009). Microsatellite variation and population genetic structure of a neotropical endangered Bryconinae species *Brycon insignis* Steindachner, 1877: Implications for its conservation and sustainable management. *Neotrop. Ichthyol.* **7,** 395–402.

Matsuoka, Y., Vigouroux, Y., Goodman, M. M., Sanchez, G. J., Buckler, E., and Doebley, J. (2002). A single domestication for maize shown by multilocus microsatellite genotyping. *Proc. Natl. Acad. Sci. U.S.A.* **99,** 6080–6084.

Mattiello, L., Kirst, M., da Silva, F. R., Jorge, R. A., and Menossi, M. (2010). Transcriptional profile of maize roots under acid soil growth. *BMC Plant Biol.* **10,** 196.

Mattsson, J., Riyal, D., and Foster, A. (2007). The application of TILLING technology in the breeding of hybrid Poplar. *Plant & Animal Genomes XV Conference, January 13-17, 2007.* .

Maukonen, J., Mättö, J., Wirtanen, G., Raaska, L., Mattila-Sandholm, T., and Saarela, M. (2003). Methodologies for the characterization of microbes in industrial environments: A review. *J. Ind. Microbiol. Biotechnol.* **30,** 327–356.

Maxime, V. (2008). The physiology of triploid fish: Current knowledge and comparisons with diploid fish. *Fish Fish.* **9,** 67–78.

Maxted, N., and Kell, S. P. (2009). Establishment of a global network for the *in situ* conservation of crop wild relatives: Status and needs. *Background Study Paper No.* 39. ftp://ftp.fao.org/docrep/fao/meeting/017/ak570e.pdf.

Maxwell, W. M. C., and Evans, G. (2009). Current status and future prospects for reproductive technologies in small ruminants. *Proc. Assoc. Advmt. Anim. Breed. Genet.* **18,** 287–295.

Mba, C., Afza, R., Jankowicz-Cieslak, J., Bado, S., Matijevic, M., Huynh, O., and Till, B. J. (2009). Enhancing genetic diversity through induced mutagenesis in vegetatively propagated plants. *In* "Induced Plant Mutations in the Genomics era, Proceedings of an International Joint FAO/IAEA Symposium, 2008" (Q. Y. Shu, ed.), pp. 262–265. FAO, Rome.

Mbogoh, S. G., Wambugu, F. M., and Wakhusama, S. (2003). Socio-economic impact of biotechnology applications: Some lessons from the pilot tissue culture (tc) banana production promotion project in Kenya, 1997-2002. Proceedings of the 25th International Conference of Agricultural Economists (IAAE). pp. 1084–1094. http://www.ecsocman.edu.ru/data/302/661/1219/089.pdf.

McAndrew, B., and Napier, J. (2010). Application of genetics and genomics to aquaculture development: Current and future directions. *J. Agric. Sci.* **149,** 143–151.

McCallum, C. M., Comai, L., Greene, E. A., and Henikoff, S. (2000). Targeting induced local lesions IN genomes (TILLING) for plant functional genomics. *Plant Physiol.* **123,** 439–442.

McCarthy, S. D., Butler, S. T., Patton, J., Daly, M., Morris, D. G., Kenny, D. A., and Waters, S. M. (2009). Differences in the expression of genes involved in the somatotropic axis in divergent strains of Holstein-Friesian dairy cows during early and mid lactation. *J. Dairy Sci.* **92,** 5229–5238.

McCarthy, F. M., Gresham, C. R., Buza, T. J., Chouvarine, P., Pillai, L. R., Kumar, R., Ozkan, S., Wang, H., Manda, P., Arick, T., Bridges, S. M., and Burgess, S. C. (2011). AgBase: Supporting functional modeling in agricultural organisms. *Nucleic Acids Res.* **39,** D497–D506.

McCusker, M. R., Paterson, I. G., and Bentzen, P. (2008). Microsatellite markers discriminate three species of North Atlantic wolffishes (*Anarhichas* spp.). *J. Fish Biol.* **72,** 375–385.

Medina, M., and Sachs, J. L. (2010). Symbiont genomics, our new tangled bank. *Genomics* **95,** 129–137.

Meeusen, E. N., Walker, J., Peters, A., Pastoret, P. P., and Jungersen, G. (2007). Current status of veterinary vaccines. *Clin. Microbiol. Rev.* **20,** 489–510.

Mendoza, L. M., Merín, M. G., Morata, V. I., and Farías, M. E. (2011). Characterization of wines produced by mixed culture of autochthonous yeasts and *Oenococcus oeni* from the northwest region of Argentina. *J. Ind. Microbiol. Biotechnol.* **38,** 1777–1785.

Menegatti, A. C., Tavares, C. P., Vernal, J., Klein, C. S., Huergo, L., and Terenzi, H. (2010). First partial proteome of the poultry pathogen *Mycoplasma synoviae*. *Vet. Microbiol.* **145,** 134–141.

Meuwissen, T. H. E., Hayes, B. J., and Goddard, M. E. (2001). Prediction of total genetic value using genome-wide dense marker maps. *Genetics* **157,** 1819–1929.

Mia, M. A. B., and Shamsuddin, Z. H. (2010). Rhizobium as a crop enhancer and biofertilizer for increased cereal production. *Afr. J. Biotechnol.* **9,** 6001–6009.

Mian, M. H. (2002). Azobiofer: A technology of production and use of *Azolla* as biofertiliser for irrigated rice and fish cultivation. *In* "Biofertilisers in Action" (I. R. Kennedy and A. T. M. A. Choudhury, eds.), pp. 45–54. Rural Industries Research and Development Corporation, Canberra.

Miller, A. J., and Schaal, B. A. (2006). Domestication and the distribution of genetic variation in wild and cultivated populations of the Mesoamerican fruit tree *Spondias purpurea* L. (Anacardiaceae). *Mol. Ecol.* **15,** 1467–1480.

Miller, S. A., Beed, F. D., and Harmon, C. L. (2009). Plant disease diagnostic capabilities and networks. *Annu. Rev. Phytopathol.* **47,** 15–38.

Miller, K. M., Li, S., Kaukinen, K. H., Ginther, N., Hammill, E., Curtis, J. M. R., Patterson, D. A., Sierocinski, T., Donnison, L., Pavlidis, P., *et al.* (2011). Genomic signatures predict migration and spawning failure in wild Canadian salmon. *Science* **331,** 214–217.

Milus, E. A., Kristensen, K., and Hovmøller, M. S. (2009). Evidence for increased aggressiveness in a recent widespread strain of *Puccinia striiformis* f. sp. tritici causing stripe rust of wheat. *Phytopathology* **99,** 89–94.

Miranda, L. A., Cassará, M. C., and Somoza, G. M. (2005). Increase in milt production by hormonal treatment in the pejerrey fish *Odontesthes bonariensis* (Valenciennes 1835). *Aquacult. Res.* **36,** 1473–1479.

Mocali, S., and Benedetti, A. (2010). Exploring research frontiers in microbiology: The challenge of metagenomics in soil microbiology. *Vet. Immunol. Immunopathol.* **138,** 280–291.

Mochida, K., and Shinozaki, K. (2010). Genomics and bioinformatics resources for crop improvement. *Plant Cell Physiol.* **51,** 497–523.

Moen, T., Hayes, B., Baranski, M., Berg, P. R., Kjoglum, S., Koop, B. F., Davidson, W. S., Omholt, S. W., and Lien, S. (2008). A linkage map of the Atlantic salmon (*Salmo salar*) based on EST-derived SNP markers. *BMC Genomics* **9,** 223.

Moen, T., Baranski, A., Sonesson, A. K., and Kjoeglum, S. (2009). Confirmation and fine mapping of a major QTL for resistance to infectious pancreatic necrosis in Atlantic salmon (*Salmo salar*): Population level association between markers and trait. *BMC Genomics* **10,** 368.

Morrell, P., and Clegg, M. (2007). Genetic evidence for a second domestication of barley (*Hordeum vulgare*) east of the Fertile Crescent. *Proc. Natl. Acad. Sci. U.S.A.* **104,** 3289–3294.

Morrell, P. L., Williams-Coplin, T. D., Lattu, A. L., Bowers, J. E., Chandler, J. M., and Paterson, A. H. (2005). Crop-to-weed introgression has impacted allelic composition of johnsongrass populations with and without recent exposure to cultivated sorghum. *Mol. Ecol.* **14,** 2143–2154.

Morris, M., Dreher, K., Ribaut, J. M., and Khairallah, M. (2003). Money matters (II): Costs of maize inbred line conversion schemes at CIMMYT using conventional and marker-assisted selection. *Mol. Breed.* **11,** 235–247.

Morrissey, M. B., and Ferguson, M. M. (2011). Individual variation in movement throughout the life cycle of a stream-dwelling salmonid fish. *Mol. Ecol.* **20,** 235–248.

Motsara, M. R., and Roy, R. N. (2008). Guide to laboratory establishment for plant nutrient analysis. FAO Fertilizer and Plant Nutrition Bulletin 19. www.fao.org/docrep/011/i0131e/i0131e00.htm.

Moukadiri, O., Lopes, C. R., and Cornejo, M. J. (1999). Physiological and genomic variations in rice cells recovered from direct immersion and storage in liquid nitrogen. *Physiol. Plant.* **105,** 442–449.

Mountzouris, K. C., Tsistsikos, P., Kalamara, E., Nitsh, S., Schatzmayr, G., and Fegeros, K. (2007). Evaluation of the efficacy of a probiotic containing *Lactobacillus*, *Bifidobacterium*, *Enterococcus*, and *Pediococcus* strains in promoting broiler performance and modulting cecal microflora composition and metabolic activities. *Poult. Sci.* **86,** 309–317.

Mtileni, B. J., Muchadeyi, F. C., Maiwashe, A., Groeneveld, E., Groeneveld, L. F., Dzama, K., and Weigend, S. (2011). Genetic diversity and conservation of South African indigenous chicken populations. *J. Anim. Breed. Genet.* **128,** 209–218.

Muchugi, A., Kadu, C., Kindt, R., Kipruto, H., Lemurt, S., Olale, K., Nyadoi, P., Dawson, I., and Jamnadass, R. (2008a). Molecular Markers for Tropical Trees: A Practical Guide to Principles and Procedures. ICRAF Technical Manual No. 9. The World Agroforestry Centre, Nairobi, Kenya.

Muchugi, A., Muluvi, G., Kindt, R., Kadu, C., Simons, A., and Jamnadass, R. (2008b). Genetic structuring of important medicinal species of genus *Warburgia* as revealed by AFLP analysis. *Tree Genet. Genomes* **4,** 787–795.

Mukhopadhyay, A., and Mukhopadhyay, U. K. (2007). Novel multiplex PCR approaches for the simultaneous detection of human pathogens: *Escherichia coli* 0157:H7 and *Listeria monocytogenes*. *J. Microbiol. Methods* **68,** 193–200.

Mumford, R., Boonham, N., Tomlinson, J., and Barker, I. (2006). Advances in molecular phytodiagnostics—New solutions for old problems. *Eur. J. Plant Pathol.* **116,** 1–19.

Muralidharan, E. M., and Kallarackal, J. (2005). Current trends in forest tree biotechnology. *In* "Plant Biotechnology and Molecular Markers" (P. S. Srivastava, A. Narula, and S. Srivastava, eds.), pp. 169–182. Kluwer Academic Publishers, The Netherlands.

Murphy, D. J. (2007). Plant Breeding and Biotechnology: Societal Context and the Future of Agriculture. Cambridge University Press, Cambridge.

Mutandwa, E. (2008). Performance of tissue-cultured sweet potatoes among smallholder farmers in Zimbabwe. *AgBioForum* **11,** 48–57.

Muyzer, G. (1999). Genetic fingerprinting of microbial communities—Present status and future perspectives. *In* "Microbial Biosystems: New Frontiers, Proceedings of the 8th International Symposium on Microbial Ecology" (C. R. Bell, M. Brylinsky, and P. Johnson-Green, eds.), pp. 1–10. Atlantic Canada Society for Microbiology, Kentville.

Mylonas, C. C., and Zohar, Y. (2007). Promoting oocyte maturation, ovulation and spawning in farmed fish. *In* "The Fish Oocyte: From Basic Studies to Biotechnological Applications" (P. J. Babin, J. Cerdà, and E. Lubzens, eds.), pp. 437–474. Springer, Dordrecht.

Mylonas, C. C., Fostier, A., and Zanuy, S. (2010). Broodstock management and hormonal manipulations of fish reproduction. *Gen. Comp. Endocrinol.* **165,** 516–534.

NACA/FAO (2011). Quarterly aquatic animal disease report (Asia and Pacific Region), 2010/3. July—September 2010. NACA, Bangkok, Thailand. http://library.enaca.org/Health/QAAD/qaad-2010-3.pdf.

Nagashima, H., Hiruma, K., Saito, H., Tomii, R., Ueno, S., Nakayama, N., Matsunari, H., and Kurome, M. (2007). Production of live piglets following cryopreservation of embryos derived from *in vitro*-matured oocytes. *Biol. Reprod.* **76,** 900–905.

Nagata, T., and Bajaj, Y. P. S. (2001). Biotechnology in Agriculture and Forestry 49: Somatic Hybridization in Crop Improvement II. Springer.

Naik, P. S., and Karihaloo, J. L. (2007). Micropropagation for Production of Quality Potato Seed in Asia Pacific. Asia-Pacific Consortium on Agricultural Biotechnology, New Delhi.

Nam, Y. K., and Kim, D. S. (2004). Ploidy status of progeny from the crosses between tetraploid males and diploid females in mud loach (*Misgurnus mizolepis*). *Aquaculture* **236,** 575–582.

Na-Nakorn, U., Rangsin, W., and Boon-ngam, J. (2004). Allotriploidy increases sterility in the hybrid between *Clarias macrocephalus* and *Clarias gariepinus*. *Aquaculture* **237,** 73–88.

Narayanan, C., Wali, S. A., Shukla, N., Kumar, R., Mandal, A. K., and Ansari, S. A. (2007). RAPD and ISSR markers for molecular characterization of teak (*Tectona grandis*) plus trees. *J. Trop. Forest Sci.* **19,** 218–225.

Narsai, R., Castleden, I., and Whelan, J. (2010). Common and distinct organ and stress responsive transcriptomic patterns in *Oryza sativa* and *Arabidopsis thaliana*. *BMC Plant Biol.* **10,** 262.

Nayak, S. K. (2010a). Role of gastrointestinal microbiota in fish. *Aquacult. Res.* **41,** 1553–1573.

Nayak, S. K. (2010b). Probiotics and immunity: A fish perspective. *Fish Shellfish Immunol.* **29,** 2–14.

Neale, D. B., and Ingvarsson, P. K. (2008). Population, quantitative and comparative genomics of adaptation in forest trees. *Curr. Opin. Plant Biol.* **11,** 149–155.

Neale, D. B., and Kremer, A. (2011). Forest tree genomics: Growing resources and applications. *Nat. Rev. Genet.* **12,** 111–122.

Negri, V., and Tiranti, B. (2010). Effectiveness of *in situ* and *ex situ* conservation of crop diversity. What a *Phaseolus vulgaris* L. landrace case study can tell us. *Genetica* **138,** 985–998.

Nehra, N. S., Becwar, M. R., Rottmann, W. H., Pearson, L., Chowdhury, K., Chang, S. J., Wilde, H. D., Kodrzycki, R. J., Zhang, C. S., Gause, K. C., *et al.* (2005). Forest biotechnology: Innovative methods, emerging opportunities. *In Vitro Cell Dev. Biol. Plant* **41,** 701–717.

Nell, J. A. (2002). Farming triploid oysters. *Aquaculture* **210,** 69–88.

Neumann, K., Kobiljski, B., Dencic, S., Varshney, R. K., and Borner, A. (2011). Genome-wide association mapping: A case study in bread wheat (*Triticum aestivum* L.). *Mol. Breed.* **27,** 37–58.

Nguyen, T. T. T., Hurwood, D., Mather, P., Na-Nakorn, N., Kamonrat, W., and Bartley, D. (2006a). Manual on Applications of Molecular Tools in Aquaculture and Inland Fisheries Management, Part 1: Conceptual Basis of Population Genetic Approaches. NACA, Bangkok.

Nguyen, T. T. T., Hurwood, D., Mather, P., Na-Nakorn, N., Kamonrat, W., and Bartley, D. (2006b). Manual on Applications of Molecular Tools in Aquaculture and Inland Fisheries Management, Part 2: Laboratory Protocols and Data Analysis. NACA, Bangkok.

Nicolaisen, M., Justesen, A. F., Thrane, U., Skouboe, P., and Holmstrom, K. (2005). An oligonucleotide microarray for the identification and differentiation of trichothecene producing and non-producing *Fusarium* species occurring on cereal grain. *J. Microbiol. Methods* **62,** 57–69.

Nielsen, E. E., Hemmer-Hansen, J., Poulsen, N. A., Loeschcke, V., Moen, T., Johansen, T., Mittelholzer, C., Taranger, G.-L., Ogden, R., and Carvalho, G. R. (2009). Genomic signatures of local directional selection in a high gene flow marine organism; the Atlantic cod (*Gadus morhua*). *BMC Evol. Biol.* **9,** 276.

Nimbkar, C., Ghalsasi, P. M., Nimbkar, B. V., Walkden-Brown, S. W., Maddox, J. F., Gupta, V. S., Pardeshi, V. C., Ghalsaasi, P. P., and van der Werf, J. H. J. (2007). Reproductive performance of Indian crossbred Deccani ewes carrying the *FecB* mutation. *Proc. Assoc. Adv. Anim. Br. Genet.* **17,** 430–433.

Nordborg, M., and Weigel, D. (2008). Next-generation genetics in plants. *Nature* **456,** 720–723.

Norris, A. T., Bradley, D. G., and Cunningham, E. P. (2000). Parentage and relatedness determination in farmed Atlantic salmon (*Salmo salar*) using microsatellite markers. *Aquaculture* **182,** 73–83.

Novaes, E., Drost, D. R., Farmerie, W. G., Pappas, G. J., Jr., Grattapaglia, D., Sederoff, R. R., and Kirst, M. (2008). High-throughput gene and SNP discovery in *Eucalyptus grandis*, an uncharacterized genome. *BMC Genomics* **9,** 312.

Odame, H. (2002). Smallholder access to biotechnology: Case of Rhizobium inocula in Kenya. *Econ. Pol. Wkly* **37,** 2748–2755.

Ogden, R. (2008). Fisheries forensics: The use of DNA tools for improving compliance, traceability and enforcement in the fishing industry. *Fish Fish.* **9,** 462–472.

Ohsawa, T., and Ide, Y. (2008). Global patterns of genetic variation in plant species along vertical and horizontal gradients on mountains. *Global Ecol. Biogeogr.* **17,** 152–163.

OIE (2009). Biotechnology in the diagnosis of infectious diseases and vaccine development. Chapter 1.1.7. in the Manual of Diagnostic Tests and Vaccines for Terrestrial Animals 2009. http://web.oie.int/eng/normes/mmanual/2008/pdf/1.1.07_BIOTECHNOLOGY.pdf.

OIE (2010a). The application of biotechnology to the development of veterinary vaccines. Chapter 1.1.7A. in the Manual of Diagnostic Tests and Vaccines for Terrestrial Animals 2010. http://web.oie.int/eng/normes/mmanual/2008/pdf/1.01.07a_VACCINES_NEW_TECH.pdf.

OIE (2010b). Manual of Diagnostic Tests and Vaccines for Terrestrial Animals 2010. http://www.oie.int/eng/normes/mmanual/a_summry.htm.

OIE (2010c). Aquatic Animal Health Code 2010. http://www.oie.int/en/international-standard-setting/aquatic-code/access-online/.

OIE (2010d). Manual of Diagnostic Tests for Aquatic Animals 2010. http://www.oie.int/international-standard-setting/aquatic-manual/access-online/.

Okumus, I., and Ciftci, Y. (2003). Fish population genetics and molecular markers: II. Molecular markers and their applications in fisheries and aquaculture. *Turk. J. Fish. Aquat. Sci.* **3,** 51–79.

Okutsu, T., Shikina, S., Kanno, M., Takeuchi, Y., and Yoshizaki, G. (2007). Production of trout offspring from triploid salmon parents. *Science* **317,** 1517.

Oleksiak, M. F. (2010). Genomic approaches with natural fish populations. *J. Fish Biol.* **76,** 1067–1093.

Olempska-Beer, Z. S., Merker, R. I., Ditto, M. D., and DiNovi, M. J. (2006). Food-processing enzymes from recombinant microorganisms—A review. *Regul. Toxicol. Pharmacol.* **45,** 144–158.

Oliveira, C., Foresti, F., and Hilsdorf, A. W. S. (2009). Genetics of neotropical fish: From chromosomes to populations. *Fish Physiol. Biochem.* **35,** 81–100.

Orabi, J., Jahoor, A., and Backes, G. (2009). Genetic diversity and population structure of wild and cultivated barley from West Asia and North Africa. *Plant Breed.* **128,** 606–614.

Ouahmane, L., Hafidi, M., Thioulouse, J., Ducousso, M., Kisa, M., Prin, Y., Galiana, A., Boumezzough, A., and Duponnois, R. (2007). Improvement of *Cupressus atlantica* Gaussen growth by inoculation with native arbuscular mycorrhizal fungi. *J. Appl. Microbiol.* **103,** 683–690.

Oumar, I., Mariac, C., Pham, J. L., and Vigouroux, Y. (2008). Phylogeny and origin of pearl millet (*Pennisetum glaucum* [L.] R. Br.) as revealed by microsatellite loci. *Theor. Appl. Genet.* **117,** 489–497.

Owiny, O. D., Barry, D. M., Agaba, M., and Godke, R. A. (2009). *In vitro* production of cattle × buffalo hybrid embryos using cattle oocytes and African buffalo (*Syncerus caffer* caffer) epididymal sperm. *Theriogenology* **71,** 884–894.

Ozkan, E., Soysal, M. I., Ozder, M., Koban, E., Sahin, O., and Togan, I. (2009). Evaluation of parentage testing in the Turkish Holstein population based on 12 microsatellite loci. *Livest. Sci.* **124**(1–3), 101–106.

Paiva, J. A. P., Garcés, M., Alves, A., Garnier-Géré, P., Rodrigues, J. C., Lalanne, C., Porcon, S., Le Provost, G., da Silva Perez, D., Brach, J., *et al.* (2008). Molecular and phenotypic profiling from the base to the crown in maritime pine wood-forming tissue. *New Phytol.* **178,** 283–301.

Paiva, J. A. P., Prat, E., Vautrin, S., Santos, M. D., San-Clemente, H., Brommonschenkel, S., Fonseca, P. G. S., Grattapaglia, D., Song, X., Ammiraju, J. S. S., *et al.* (2011). Advancing *Eucalyptus* genomics: Identification and sequencing of lignin biosynthesis genes from deep-coverage BAC libraries. *BMC Genomics* **12,** 137.

Pakkad, G., Ueno, S., and Yoshimaru, H. (2008). Gene flow pattern and mating system in a small population of *Quercus semiserrata* Roxb. (Fagaceae). *For. Ecol. Manage.* **255,** 3819–3826.

Pan, G., and Yang, J. (2010). Analysis of microsatellite DNA markers reveals no genetic differentiation between wild and hatchery populations of Pacific threadfin in Hawaii. *Int. J. Biol. Sci.* **6,** 827–833.

Panda, A. K., Rama Rao, S. S., Raju, M. V. L. N., and Sharma, S. S. (2008). Effect of probiotic (*Lactobacillus sporogenes*) feeding on egg production and quality, yolk cholesterol and humoral immune response of white leghorn layer breeders. *J. Sci. Food Agric.* **88,** 43–47.

Pandian, T. J., and Kirankumar, S. (2003). Androgenesis and conservation of fishes. *Curr. Sci.* **85,** 917–931.

Paniagua-Chavez, C. G., and Tiersch, T. R. (2001). Laboratory studies of cryopreservation of sperm and trochophore larvae of the eastern oyster. *Cryobiology* **43,** 211–223.

Panigrahi, A., and Azad, I. S. (2007). Microbial intervention for better fish health in aquaculture: The Indian scenario. *Fish Physiol. Biochem.* **33,** 429–440.

Panis, B., and Lambardi, M. (2006). Status of cryopreservation technologies in plants (crops and forest trees). *In* "The Role of Biotechnology in Exploring and Protecting Agricultural Genetic Resources" (J. Ruane and A. Sonnino, eds.), pp. 61–78. FAO, Rome.

Panis, B., Van den Houwe, I., Piette, B., and Swennen, R. (2007). Cryopreservation of the banana germplasm collection at the ITC (INIBAP Transit Centre). *Proc. 1st Meeting of COST 871 Working Group 2: Technology, application and validation of plant cryopreservation. Florence, Italy, 10–13 May, 2007. pp. 34–35.*

Paoli, P. D. (2005). Biobanking in microbiology: From sample collection to epidemiology, diagnosis and research. *FEMS Microbiol. Rev.* **29,** 897–910.

Papa, R. (2005). Gene flow and introgression between domesticated crops and their wild relatives. *Paper at "The role of biotechnology for the characterization and conservation of crop, forestry, animal and fishery genetic resources". International Workshop.* pp. 5–7. Turin, Italy. March 2005. http://www.fao.org/biotech/docs/papa.pdf.

Papa, R., and Gepts, P. (2003). Asymmetry of gene flow and differential geographical structure of molecular diversity in wild and domesticated common bean (*Phaseolus vulgaris* L.) from Mesoamerica. *Theor. Appl. Genet.* **106,** 239–250.

Parchman, T. L., Geist, K. S., Grahnen, J. A., Benkman, C. W., and Buerkle, C. A. (2010). Transcriptome sequencing in an ecologically important tree species: Assembly, annotation, and marker discovery. *BMC Genomics* **11,** 180.

Park, Y. (2002). Implementation of conifer somatic embryogenesis in clonal forestry: Technical requirements and deployment considerations. *Ann. For. Sci.* **59,** 651–656.

Park, Y. S., and Klimaszewska, D. (2003). Achievements and challenges in conifer somatic embryogenesis for clonal forestry. XII World Forestry Conference, 2003, Quebec City, Canada. http://www.fao.org/DOCREP/ARTICLE/WFC/XII/0221-B2.HTM.

Park, S., Keathley, D. E., and Han, K. H. (2008). Transcriptional profiles of the annual growth cycle in *Populus deltoides. Tree Physiol.* **28,** 321–329.

Parkouda, C., Nielsen, D. S., Azokpota, P., Ouoba, L. I. I., Amoa-Awua, W. K., Thorsen, L., Hounhouigan, J. D., Jensen, J. S., Tano-Debrah, K., Diawara, B., *et al.* (2009). The microbiology of alkaline-fermentation of indigenous seeds used as food condiments in Africa and Asia. *Crit. Rev. Microbiol.* **35,** 139–156.

Parry, M. A., Madgwick, P. J., Bayon, C., Tearall, K., Hernandez-Lopez, A., Baudo, M., Rakszegi, M., Hamada, W., Al-Yassin, A., Ouabbou, H., *et al.* (2009). Mutation discovery for crop improvement. *J. Exp. Bot.* **60,** 2817–2825.

Patel, S., and Goyal, A. (2010). Isolation, characterization and mutagenesis of exopolysaccharide synthesizing new strains of lactic acid bacteria. *Internet J. Microbiol.* **8,** 3–4.

Pauk, J., Jancsó, M., and Simon-Kiss, I. (2009). Rice doubled haploids and breeding. *In* "Advances in Haploid Production in Higher Plants" (A. Touraev, B. P. Forster, and S. M. Jain, eds.), pp. 189–197. Springer.

Paynter, S., Cooper, A., Thomas, N., and Fuller, B. (1997). Cryopreservation of multicellular embryos and reproductive tissues. *In* "Reproductive Tissue Banking: Scientific Principles" (A. M. Karow and J. K. Critser, eds.), pp. 359–397. Academic Press, San Diego.

Peatman, E., Baoprasertkul, P., Terhune, J., Xu, P., Nandi, S., Kucuktas, H., Li, P., Wang, S., Somridhivej, B., Dunham, R., *et al.* (2007). Expression analysis of the acute phase response in channel catfish (*Ictalurus punctatus*) after infection with a Gram-negative bacterium. *Dev. Comp. Immunol.* **31,** 1183–1196.

Peck, L. S., Clark, M. S., Clarke, A., Cockell, C. S., Convey, P., Detrich, H. W., III, Fraser, K. P. P., Johnston, I. A., Methe, B. A., Murray, A. E., *et al.* (2005). Genomics: Applications to Antarctic ecosystems. *Polar Biol.* **28,** 351–365.

Pei, Z., Gao, J., Chen, Q., Wei, J., Li, Z., Luo, F., Shi, L., Ding, B., and Sun, S. (2010). Genetic diversity of elite sweet sorghum genotypes assessed by SSR markers. *Biol. Plant.* **54,** 653–658.

Peinado-Guevara, L. I., and López-Meyer, M. (2006). Detailed monitoring of white spot syndrome virus (WSSV) in shrimp commercial ponds in Sinaloa, Mexico by nested PCR. *Aquaculture* **25,** 33–45.

Pelgas, B., Beauseigle, S., Achere, V., Jeandroz, S., Bousquet, J., and Isabel, N. (2006). Comparative genome mapping among *Picea glauca*, *P. mariana* X *P. rubens* and *P. abies*, and correspondence with other Pinaceae. *Theor. Appl. Genet.* **113,** 1371–1393.

Pereira, R. M., da Silveira, E. L., Scaquitto, D. C., Pedrinho, E. A. N., Val-Moraes, S. P., Wickert, E., Carareto-Alves, L. M., and de Macedo Lemos, E. G. (2006). Molecular characterization of bacterial populations of different soils. *Braz. J. Microbiol.* **37,** 439–447.

Pereira, J. J., Shanmugam, S. A., Sulthana, M., and Sundaraj, V. (2009). Effect of vaccination on vibriosis resistance of *Fenneropenaeus indicus*. *Tamil Nadu J. Vet. Anim. Sci.* **5,** 246–250.

Pereira, J. C., Lino, P. G., Leitão, A., Joaquim, S., Chaves, R., Pousão-Ferreira, P., Guedes-Pinto, H., and dos Santos, M. N. (2010). Genetic differences between wild and hatchery populations of *Diplodus sargus* and *D. vulgaris* inferred from RAPD markers: Implications for production and restocking programs design. *J. Appl. Genet.* **51,** 67–72.

Perera, L., Russell, J. R., Provan, J., McNicol, J. W., and Powell, W. (1998). Evaluating genetic relationships between indigenous coconut (*Cocos nucifera* L.) accessions from Sri Lanka by means of AFLP profiling. *Theor. Appl. Genet.* **96,** 545–550.

Perry, L., Heard, P., Kane, M., Kim, H., Savikhin, S., Dominguez, W., and Applegate, B. (2007). Application of multiplex polymerase chain reaction to the detection of pathogens in food. *J. Rapid Methods Autom. Microbiol.* **15,** 176–198.

Pessoa-Filho, M., Rangel, P. H., and Ferreira, M. E. (2010). Extracting samples of high diversity from thematic collections of large gene banks using a genetic-distance based approach. *BMC Plant Biol.* **10,** 127–136.

Peter, C., Bruford, M., Perez, T., Dalamitra, S., Hewitt, G., Erhardt, G., and the ECONOGENE Consortium (2007). Genetic diversity and subdivision of 57 European and Middle-Eastern sheep breeds. *Anim. Genet.* **38,** 37–44.

Peternel, S., Gabrovsek, K., Gogala, N., and Regvar, M. (2009). *In vitro* propagation of European aspen (*Populus tremula* L.) from axillary buds via organogenesis. *Sci. Hortic.* **121,** 109–112.

Petit, R. J., and Hampe, A. (2006). Some evolutionary consequences of being a tree. *Annu. Rev. Ecol. Evol. Syst.* **37,** 187–214.

Picchietti, S., Fausto, A. M., Randelli, E., Carnevali, O., Taddei, A. R., Buonocore, F., Scapigliati, G., and Abelli, L. (2009). Early treatment with *Lactobacillus delbrueckii* strain induces an increase in intestinal T-cells and granulocytes and modulates immunerelated genes of larval *Dicentrarchus labrax* (L.). *Fish Shellfish Immunol.* **26,** 368–376.

Piferrer, F. (2001). Endocrine sex control strategies for the feminization of teleost fish. *Aquaculture* **197,** 229–281.

Piferrer, F., Beaumont, A. R., Falguiere, J.-C., and Colombo, L. (2006). Performance improvements by polyploidization in aquaculture. *In* "Genetic Impact of Aquaculture Activities on Native Populations" (T. Svåsand, D. Crosetti, E. García-Vázquez, and E. Verspoor, eds.), pp. 100–103. Genimpact final scientific report.

Piferrer, F., Beaumont, A., Falguière, J. C., Flajšhans, M., Haffray, P., and Colombo, L. (2009). Polyploid fish and shellfish: Production, biology and applications to aquaculture for performance improvement and genetic containment. *Aquaculture* **293,** 125–156.

Piggott, M. P., Chao, N. L., and Beheregaray, L. B. (2011). Three fishes in one: Cryptic species in an Amazonian floodplain forest specialist. *Biol. J. Linn. Soc.* **102,** 391–403.

Pijut, P. M., Woeste, K. E., Vengadesan, G., and Michler, C. H. (2007). Technological advances in temperate hardwood tree improvement including breeding and molecular marker applications. *In Vitro Cell. Dev. Biol. Plant* **43,** 283–303.

Pijut, P. M., Lawson, S. S., and Michler, C. H. (2011). Biotechnological efforts for preserving and enhancing temperate hardwood tree biodiversity, health, and productivity. *In Vitro Cell. Dev. Biol. Plant* **47,** 123–147.

Pineda, A., Zheng, S. J., van Loon, J. J., Pieterse, C. M., and Dicke, M. (2010). Helping plants to deal with insects: The role of beneficial soil-borne microbes. *Trends Plant Sci.* **15,** 507–514.

Pinto, P. M., Klein, C. S., Zaha, A., and Ferreira, H. B. (2009). Comparative proteomic analysis of pathogenic and non-pathogenic strains from the swine pathogen *Mycoplasma hyopneumoniae*. *Proteome Sci.* **7,** 45.

Pípalová, I. (2006). A review of grass carp use for aquatic weed control and its impact on water bodies. *J. Aquat. Plant Manag.* **44,** 1–12.

Planes, S., Jones, G. P., and Thorrold, S. R. (2009). Larval dispersal connects fish populations in a network of marine protected areas. *Proc. Natl. Acad. Sci. U.S.A.* **106,** 5693–5697.

Porwal, S., Lal, S., Cheema, S., and Kalia, V. C. (2009). Phylogeny in aid of the present and novel microbial lineages: Diversity in *Bacillus*. *PLoS One* **4,** e4438.

Potter, A., Gerdts, V., and Littel-van den Hurk, S. D. (2008). Veterinary vaccines: Alternatives to antibiotics? *Anim. Health Res. Rev.* **9,** 187–199.

Poulos, B. T., Tang-Nelson, K. F., Pantoja, C. R., Nunan, L. M., Navarro, S. A., Redman, R. M., and Mohney, L. L. (2006). Application of molecular diagnostic methods to penaeid shrimp diseases: Advances of the past 10 years for control of viral diseases in farmed shrimp. *Dev. Biol. (Basel)* **126,** 117–122.

Povh, J. A., Lopera-Barrero, N. M., Ribeiro, R. P., Lupchinski, E., Jr., Gomes, P. C., and Lopes, T. S. (2008). Genetic monitoring of fish repopulation programs using molecular markers. *Cien. Inv. Agr.* **35,** 1–10.

Prado, S., Romalde, J. L., and Barja, J. L. (2010). Review of probiotics for use in bivalve hatcheries. *Vet. Microbiol.* **145,** 187–197.

Prasanna, R., Singh, R. N., Joshi, M., Madhan, K., Pal, R. K., and Nain, L. (2011). Monitoring the biofertilizing potential and establishment of inoculated cyanobacteria in soil using physiological and molecular markers. *J. Appl. Phycol.* **23,** 301–308.

Prentice, J. R., and Anzar, M. (2011). Cryopreservation of mammalian oocyte for conservation of animal genetics. *Vet. Med. Int.* **2011,** 146405.

Presti, R. L., Lisa, C., and Di Stasio, L. (2009). Molecular genetics in aquaculture. *Ital. J. Anim. Sci.* **8,** 299–313.

Primmer, C. (2006). Genetic characterization of populations and its use in conservation decision-making in fish. *In* "The Role of Biotechnology in Exploring and Protecting Agricultural Genetic Resources" (J. Ruane and A. Sonnino, eds.), pp. 97–104. FAO, Rome.

Purugganan, M. D., and Fuller, D. Q. (2009). The nature of selection during plant domestication. *Nature* **457,** 843–848.

Qanbari, S., Pimentel, E. C. G., Tetens, J., Thaller, G., Lichtner, P., Sharifi, A. R., and Simianer, H. (2010). A genome-wide scan for signatures of recent selection in Holstein cattle. *Anim. Genet.* **41,** 377–389.

Qi, Z., Zhang, X.-H., Boon, N., and Bossier, P. (2009). Probiotics in aquaculture of China—Current state, problems and prospect. *Aquaculture* **290,** 15–21.

Qi-Lun, Y., Ping, F., Ke-Cheng, K., and Guang-Tang, P. (2008). Genetic diversity based on SSR markers in maize (*Zea mays* L.) landraces from Wuling mountain region in China. *J. Genet.* **87,** 287–291.

Quirino, B. F., Candido, E. S., Campos, P. F., Franco, O. L., and Krüger, R. H. (2010). Proteomic approaches to study plant-pathogen interactions. *Phytochemistry* **71,** 351–362.

Raadsma, H. W., Thomson, P. C., Zenger, K. R., Cavanagh, C., Lam, M. K., Jonas, E., Jones, M., Attard, G., Palmer, D., and Nicholas, F. W. (2009). Mapping quantitative trait loci (QTL) in sheep. I. A new male framework linkage map and QTL for growth rate and body weight. *Genet. Sel. Evol.* **41,** 34.

Radulovici, A. E., Archambault, P., and Dufresne, F. (2010). DNA barcodes for marine biodiversity: Moving fast forward? *Diversity* **2,** 450–472.

Rantsiou, K., Alessandria, V., Urso, R., Dolci, P., and Cocolin, L. (2008). Detection, quantification and vitality of *Listeria monocytogenes* in food as determined by quantitative PCR. *Int. J. Food Microbiol.* **121,** 99–105.

Rasmussen, R. S., and Morrissey, M. T. (2007). Biotechnology in aquaculture: Transgenics and polyploidy. *Compr. Rev. Food Sci. Food Saf.* **6,** 2–16.

Rasmussen, K. K., and Kollmann, J. (2008). Low genetic diversity in small peripheral populations of a rare European tree (*Sorbus torminalis*) dominated by clonal reproduction. *Conserv. Genet.* **9,** 1533–1539.

Rast, J. P., and Messier-Solek, C. (2008). Marine invertebrate genome sequences and our evolving understanding of animal immunity. *Biol. Bull.* **214,** 274–283.

Rao, N. K. (2004). Plant genetic resources: Advancing conservation and use through biotechnology. *Afr. J. Biotechnol.* **3,** 136–145.

Rath, D. (2008). Status of sperm sexing technologies. Proceedings of the 24th AETE Meeting. pp. 89–94. PAU, France. http://www.aete.eu/pdf_publication/24.pdf.

Rath, D., and Johnson, L. A. (2008). Application and commercialization of flow cytometrically sex-sorted semen. *Reprod. Domest. Anim.* **43**(Suppl. 2), 338–346.

Ray, R. R. (2011). Microbial isoamylases: An overview. *Am. J. Food Technol.* **6,** 1–18.

Reed, B. M. (2008). Plant Cryopreservation: A Practical Guide. Springer, New York.

Reed, B. M., Engelmann, F., Dulloo, M. E., and Engels, J. M. M. (2004). Technical Guidelines for the Management of Field and In Vitro Germplasm Collections. Handbook for Genebanks N 7. IPGRI/SGRP, Rome.

Renau-Morata, B., Arrillaga, I., and Segura, J. (2006). *In vitro* storage of cedar shoot cultures under minimal growth conditions. *Plant Cell Rep.* **25,** 636–642.

Renn, S. C. P., Aubin-Horth, N., and Hofmann, H. A. (2004). Biologically meaningful expression profiling across species using heterologous hybridization to a cDNA microarray. *BMC Genomics* **5,** 42.

Ribaut, J. M., de Vicente, M. C., and Delannay, X. (2010). Molecular breeding in developing countries: Challenges and perspectives. *Curr. Opin. Plant Biol.* **13,** 213–218.

Rice, E. B., Smith, M. E., Mitchell, S. E., and Kresovich, S. (2006). Conservation and change: A comparison of *in situ* and *ex situ* conservation of Jala maize germplasm. *Crop Sci.* **46,** 428–436.

Rinaldi, M., Li, R. W., and Capuco, A. V. (2010). Mastitis associated transcriptomic disruptions in cattle. *Vet. Immunol. Immunopathol.* **138,** 267–279.

Rincón, G., Islas-Trejo, A., Casellas, J., Ronin, Y., Soller, M., Lipkin, E., and Medrano, J. F. (2009). Fine mapping and association analysis of a quantitative trait locus for milk production traits on *Bos taurus* autosome 4. *J. Dairy Sci.* **92,** 758–764.

Riyaz-Ul-Hassan, S., Verma, V., and Qazi, G. N. (2008). Evaluation of three different molecular markers for the detection of *Staphylococcus aureus* by polymerase chain reaction. *Food Microbiol.* **25,** 452–459.

Roberge, C., Páez, D. J., Rossignol, O., Guderley, H., Dodson, J., and Bernatchez, L. (2007). Genome-wide survey of the gene expression response to saprolegniasis in Atlantic salmon. *Mol. Immunol.* **44,** 1374–1383.

Robinson, A. S., Vreysen, M. J. B., Hendrichs, J., and Feldmann, U. (2009). Enabling technologies to improve area-wide integrated pest management programmes for the control of screwworms. *Med. Vet. Entomol.* **23**(Suppl. 1), 1–7.

Rodriguez-Dorta, N., Cognié, Y., González, F., Poulin, N., Guignot, F., Touzé, J., Baril, G., Cabrera, F., Álamo, D., and Batista, M. (2007). Effect of coculture with oviduct epithelial cells on viability after transfer of vitrified *in vitro* produced goat embryos. *Theriogenology* **68,** 908–913.

Rogan, D., and Babuik, L. A. (2005). Novel vaccines from biotechnology. *Rev. Sci. Tech.* **24,** 159–174.

Roger, P. (2004). N2-fixing cyanobacteria as biofertilizers in rice fields. *In* "Handbook of Microalgal Culture: Biotechnology and Applied Phycology" (A. Richmond, ed.), pp. 392–402. Blackwell Science, Oxford.

Rohrer, G. A., Freking, B. A., and Nonneman, D. (2007). Single nucleotide polymorphisms for pig identification and parentage exclusion. *Anim. Genet.* **38,** 253–258.

Rosen, G. L., Sokhansanj, B. A., Polikar, R., Bruns, M. A., Russell, J., Garbarine, E., Essinger, S., and Yok, N. (2009). Signal processing for metagenomics: Extracting information from the soup. *Curr. Genomics* **10,** 493–510.

Routray, P., Verma, D. K., Sarkar, S. K., and Sarangi, N. (2007). Recent advances in carp seed production and milt cryopreservation. *Fish Physiol. Biochem.* **33,** 413–427.

Routray, P., Dash, C., Dash, S. N., Tripathy, S., Verma, D. K., Swain, S. K., Swain, P., and Guru, B. C. (2010). Cryopreservation of isolated blastomeres and embryonic stem-like cells of *Leopard danio, Brachydanio frankei. Aquacult. Res.* **41,** 579–589.

Rowden, A., Robertson, A., Allnutt, T. R., Heredia, S., Williams-Linera, G., and Newton, A. (2004). Conservation genetics of Mexican beech, *Fagus grandifolia* Var. Mexicana. *Conserv. Genet.* **5,** 475–484.

Ruane, J., and Sonnino, A. (2006). Background document to the e-mail conference on the role of biotechnology for the characterization and conservation of crop, forest, animal and fishery genetic resources in developing countries. *In* "The Role of Biotechnology in Exploring and Protecting Agricultural Genetic Resources" (J. Ruane and A. Sonnino, eds.), pp. 151–172. FAO, Rome.

Ruane, J., Sonnino, A., and Agostini, A. (2010). Bioenergy and the potential contribution of agricultural biotechnologies in developing countries. *Biomass Bioenergy* **34,** 1427–1439.

Rudi, N., Norton, G. W., Alwang, J., and Asumugha, G. (2010). Economic impact analysis of marker-assisted breeding for resistance to pests and post harvest deterioration in cassava. *Afr. J. Agric. Resour. Econ.* **4,** 110–122.

Russell, J. R., Weber, J. C., Booth, A., Powell, W., Sotelo-Montes, C., and Dawson, I. K. (1999). Genetic variation of *Calycophyllum spruceanum* in the Peruvian Amazon Basin, revealed by amplified fragment length polymorphism (AFLP) analysis. *Mol. Ecol.* **8,** 199–204.

Rutkoski, J. E., Heffner, E. L., and Sorrells, M. E. (2011). Genomic selection for durable stem rust resistance in wheat. *Euphytica* **179,** 161–173.

Ryan, M. J., Smith, D., Bridge, P. D., and Jeffries, P. (2003). The relationship between fungal preservation method and secondary metabolite production in *Metarhizium anisopliae* and *Fusarium oxysporum. World J. Microbiol. Biotechnol.* **19,** 839–844.

Ryynanen, L., and Aronen, T. (2005). Genome fidelity during short- and long-term tissue culture and differentially cryostored meristems of silver birch (*Betula pendula*). *Plant Cell Tissue Organ Cult.* **83,** 21–32.

Saker, M. M., Youssef, S. S., Abdallah, N. A., Bashandy, H. S., and El-Sharkawy, A. M. (2005). Genetic analysis of some Egyptian rice genotypes using RAPD, SSR and AFLP. *Afr. J. Biotechnol.* **4,** 882–890.

Salonius, K., Simard, N., Harland, R., and Ulmer, J. B. (2007). The road to licensure of a DNA vaccine. *Curr. Opin. Investig. Drugs* **8,** 635–641.

Samuelsson, L. M., Björlenius, B., Förlin, L., and Larsson, D. G. J. (2011). Reproducible 1H NMR-based metabolomic responses in fish exposed to different sewage effluents in two separate studies. *Environ. Sci. Technol.* **45,** 1703–1710.

Sana, T. R., Fischer, S., Wohlgemuth, G., Katrekar, A., Jung, K.-H., Ronald, P. C., and Fiehn, O. (2010). Metabolomic and transcriptomic analysis of the rice response to the bacterial blight pathogen *Xanthomonas oryzae* pv. *Oryzae. Metabolomics* **6,** 451–465.

Sanchez, C., Martinez, M. T., Vidal, N., San-Jose, M. C., Valladares, S., and Vieitez, A. M. (2008). Preservation of *Quercus robur* germplasm by cryostorage of embryogenic cultures derived from mature trees and RAPD analysis of genetic stability. *Cryo Letters* **29,** 493–504.

Sanchez, B. C., Ralston-Hooper, K., and Sepúlveda, M. S. (2011). Review of recent proteomic applications in aquatic toxicology. *Environ. Toxicol. Chem.* **30,** 274–282.

Sanjur, O. I., Piperno, D. R., Andres, T. C., and Wessel-Beaver, L. (2002). Phylogenetic relationships among domesticated and wild species of Cucurbita (*Cucurbitaceae*) inferred from a mitochondrial gene: Implications for crop plant evolution and areas of origin. *Proc. Natl. Acad. Sci. U.S.A.* **99,** 535–540.

Sansaloni, C. P., Petroli, C. D., Carling, J., Hudson, C. J., Steane, D. A., Myburg, A. A., Grattapaglia, D., Vaillancourt, R. E., and Kilian, A. (2010). A high density diversity arrays technology (DArT) microarray for genome-wide genotyping in *Eucalyptus*. *Plant Methods* **6,** 16.

Sanscartier, D., Laing, T., Reimer, K., and Zeeb, B. (2009). Bioremediation of weathered petroleum hydrocarbon soil contamination in the Canadian High Arctic: Laboratory and field studies. *Chemosphere* **77,** 1121–1126.

Sarropoulou, E., and Fernandes, J. M. O. (2011). Comparative genomics in teleost species: Knowledge transfer by linking the genomes of model and non-model fish species. *Comp. Biochem. Physiol. Part D Genomics Proteomics* **6,** 92–102.

Savan, R., Kono, T., Itami, T., and Sakai, M. (2005). Loop-mediated isothermal amplification: An emerging technology for detection of fish and shellfish pathogens. *J. Fish Dis.* **28,** 573–581.

Savolainen, O., Pyhajarvi, T., and Knurr, T. (2007). Gene flow and local adaptation in trees. *Annu. Rev. Ecol. Evol. Syst.* **38,** 595–619.

Schaad, N. W., Gaush, P., Postnikova, E., and Frederick, R. (2001). On-site one hour PCR diagnosis of bacterial diseases. *Phytopathology* **91,** S79–S89.

Schaad, N. W., Opgenorth, D., and Gaush, P. (2002). Real-time polymerase chain reaction for one-hour on-site diagnosis of Pierce's disease of grape in early season asymptomatic vines. *Phytopathology* **92,** 721–728.

Schamberger, G. P., Ronald, L., Phillips, R. L., Jennifer, L., Jacobs, J. L., and Diez-Gonzalez, F. (2004). Reduction of *Escherichia coli* O157:H7 populations in cattle by addition of colicin E7-producing *E. coli* to feed. *Appl. Environ. Microbiol.* **70,** 6053–6060.

Schlötterer, C. (2004). The evolution of molecular markers—Just a matter of fashion? *Nat. Rev. Genet.* **5**(1), 63–69.

Schmale, D. G., III, and Munkvold, G. P. (2009). Mycotoxins in Crops: A Threat to Human and Domestic Animal Health. The American Phytopathological Society Education Center. http://www.apsnet.org/edcenter/intropp/topics/Mycotoxins/Pages/Detection.aspx.

Schneider, M. V., and Orchard, S. (2011). Omics technologies, data and bioinformatics principles. *Methods Mol. Biol.* **719,** 3–30.

Schwartz, M. W., Hoeksema, J. D., Gehring, C. A., Johnson, N. C., Klironomos, J. N., Abbott, L. K., and Pringle, A. (2006). The promise and the potential consequences of the global transport of mycorrhizal fungal inoculum. *Ecol. Lett.* **9,** 501–515.

Scocchi, A. M., and Mroginski, L. A. (2004). *In vitro* conservation of apical meristem-tip of *Melia azedarach* L. (Meliaceae) under slow-growth conditions. *Int. J. Exp. Bot.* **2004,** 137–143.

Sederoff, R., Myburg, A., and Kirst, M. (2009). Genomics, domestication, and evolution of forest trees. *Cold Spring Harb. Symp. Quant. Biol.* **74,** 303–317.

Seichter, D., Russ, I., Förster, M., and Medugorac, I. (2011). SNP-based association mapping of Arachnomelia in Fleckvieh cattle. *Anim. Genet.* **42,** 544–547.

Sellars, M. J., Li, F., Preston, N. P., and Xiang, J. (2010). Penaeid shrimp polyploidy: Global status and future direction. *Aquaculture* **310,** 1–7.

Sengun, I. Y., and Karabiyikli, S. (2011). Importance of acetic acid bacteria in food industry. *Food Control* **22,** 647–656.

Settanni, L., and Moschetti, G. (2010). Non-starter lactic acid bacteria used to improve cheese quality and provide health benefits. *Food Microbiol.* **27,** 691–697.

Setterington, E. B., and Alocilja, E. C. (2011). Rapid electrochemical detection of polyaniline-labeled *Escherichia coli* O157:H7. *Biosens. Bioelectron.* **26,** 2208–2214.

Seyedabadi, H., Amirinia, C., Banabazi, M. H., and Emrani, H. (2006). Parentage verification of Iranian Caspian horse using microsatellites markers. *Iran. J. Biotechnol.* **4,** 260–264.

Sharma, S. D. (2005). Cryopreservation of somatic embryos—An overview. *Indian J. Biotechnol.* **4,** 47–55.

Sharma, N. K., Tiwari, S. P., Tripathi, K., and Rai, A. K. (2011). Sustainability and cyanobacteria (blue-green algae): Facts and challenges. *J. Appl. Phycol.* **23,** 1059–1081.

Shepherd, M., and Williams, C. G. (2008). Comparative mapping among subsection Australes (genus *Pinus*, family Pinacaea). *Genome* **51,** 320–331.

Shepherd, L. V., Fraser, P., and Stewart, D. (2011). Metabolomics: A second-generation platform for crop and food analysis. *Bioanalysis* **3,** 1143–1159.

Shi, S., Valle-Rodríguez, J. O., Siewers, V., and Nielsen, J. (2011). Prospects for microbial biodiesel production. *Biotechnol. J.* **6,** 277–285.

Shih, T.-H., Chen, C.-H., Wang, H.-W., Pai, T.-W., and Chang, H.-T. (2010). BiMFG: Bioinformatics tools for marine and freshwater species. *J. Bioinform. Comput. Biol.* **8**(Suppl.1), 17–32.

Shikano, T., Shimada, Y., and Suzuki, H. (2008). Comparison of genetic diversity at microsatellite loci and quantitative traits in hatchery populations of Japanese flounder *Paralichthys olivaceus*. *J. Fish Biol.* **72,** 386–399.

Shirin, F., Rana, P. K., and Mandal, A. K. (2005). *In vitro* clonal propagation of mature *Tectona grandis* through axillary bud proliferation. *J. For. Res.* **10,** 465–469.

Shrestha, M. K., Volkaert, H., and Van Der Straeten, D. (2005). Assessment of genetic diversity in *Tectona grandis* using amplified fragment length polymorphism markers. *Can. J. For. Res.* **35,** 1017–1022.

Sicard, D., and Legras, J. L. (2011). Bread, beer and wine: Yeast domestication in the *Saccharomyces* sensu stricto complex. *C. R. Biol.* **334,** 229–236.

Siezen, R. J., Bayjanov, J. R., Felis, G. E., van der Sijde, M. R., Starrenburg, M., Molenaar, D., Wels, M., van Hijum, S. A., and van Hylckama Vlieg, J. E. (2011). Genome-scale diversity and niche adaptation analysis of *Lactococcus lactis* by comparative genome hybridization using multi-strain arrays. *Microb. Biotechnol.* **4,** 383–402.

Singh, J., Behal, A., Singla, N., Joshi, A., Birbian, N., Singh, S., Bali, V., and Batra, N. (2009). Metagenomics: Concept, methodology, ecological inference and recent advances. *Biotechnol. J.* **4,** 480–494.

Singh, R. K., Mishra, S. K., Singh, S. P., Mishra, N., and Sharma, M. L. (2010). Evaluation of microsatellite markers for genetic diversity analysis among sugarcane species and commercial hybrids. *Aust. J. Crop Sci.* **4,** 116–125.

Singh, H. P., Uma, S., Selvarajan, R., and Karihaloo, J. L. (2011a). Micropropagation for Production of Quality Banana Planting Material in Asia-Pacific. Asia-Pacific Consortium on Agricultural Biotechnology (APCoAB), New Delhi.

Singh, B., Kunze, G., and Satyanarayana, T. (2011b). Developments in biochemical aspects and biotechnological applications of microbial phytases. *Biotechnol. Mol. Biol. Rev.* **6,** 69–87.

Siwek, M., and Knol, E. F. (2010). Parental reconstruction in rural goat population with microsatellite markers. *Ital. J. Anim. Sci.* **9,** e50.

Skaala, O., Wennevik, V., and Glover, K. A. (2006). Evidence of temporal genetic change in wild Atlantic salmon, *Salmo salar* L., populations affected by farm escapees. *ICES J. Mar. Sci.* **63,** 1224–1233.

Skugor, S., Jørgensen, S. M., Gjerde, B., and Krasnov, A. (2009). Hepatic gene expression profiling reveals protective responses in Atlantic salmon vaccinated against furunculosis. *BMC Genomics* **10,** 503.

Skuse, G. R., and Du, C. (2008). Bioinformatics tools for plant genomics. Special issue of International Journal of Plant Genomics. Hindawi Publishing.

Smith, D., and Ryan, M. J. (2008). The impact of OECD best practice on the validation of cryopreservation techniques for microorganisms. *Cryo Letters* **29,** 63–72.

Smith, D., Ryan, M. J., and Stackebrandt, E. (2008). The *ex situ* conservation of microorganisms: Aiming at a certified quality management. *In* "Biotechnology" (H. W. Doelle and E. J. DaSilva, eds.). EOLSS Publishers, Oxford.

Sobhanian, H., Aghaei, K., and Komatsu, S. (2011). Changes in the plant proteome resulting from salt stress, toward the creation of salt-tolerant crops? *J. Proteomics* **74,** 1323–1337.

Soccol, C. R., de Souza Vandenberghe, L. P., Spier, M. R., Medeiros, A. B. P., Yamaguishi, C. T., Lindner, J. D. D., Pandey, A., and Thomaz-Soccol, V. (2010). The potential of probiotics: A review. *Food Technol. Biotechnol.* **48,** 413–434.

Sollero, B. P., Paiva, S. R., Faria, D. A., Guimarães, S. E. F., Castro, S. T. R., Egito, A. A., Albuquerque, M. S. M., Piovezan, U., Bertani, G. R., and Mariante, A.da.S. (2009). Genetic diversity of Brazilian pig breeds evidenced by microsatellite markers. *Livest. Sci.* **123,** 8–15.

Somado, E. A., Guei, R. G., and Keya, S. O. (2008). NERICA®: The New Rice for Africa—A compendium. Africa Rice Center, Cotonou.

Sommerset, I., Krossoy, B., Biering, E., and Frost, P. (2005). Vaccines for fish in aquaculture. *Expert Rev. Vaccines* **4,** 89–101.

Sonesson, A. K., and Meuwissen, T. H. E. (2009). Testing strategies for genomic selection in aquaculture breeding programs. *Genet. Sel. Evol.* **41,** 37.

Soni, K. A., Nannapaneni, R., and Tasara, T. (2011). The contribution of transcriptomic and proteomic analysis in elucidating stress adaptation responses of *Listeria monocytogenes*. *Foodborne Pathog. Dis.* **8,** 843–852.

Sonnino, A., Carena, M. J., Guimarães, E. P., Baumung, R., Pilling, D., and Rischkowsky, B. (2007). An assessment of the use of molecular markers in developing countries. *In* "Marker-Assisted Selection: Current Status and Future Perspectives in Crops, Livestock, Forestry and Fish" (E. Guimarães, J. Ruane, B. Scherf, A. Sonnino, and J. Dargie, eds.), pp. 15–26. FAO, Rome.

Sonnino, A., Dhlamini, Z., Mayer-Tasch, L., and Santucci, F. M. (2009). Assessing the socio-economic impacts of non-transgenic biotechnologies in developing countries. *In* "Socio-economic Impacts of Non-transgenic Biotechnologies in Developing Countries: The Case of Plant Micro-propagation in Africa" (A. Sonnino, Z. Dhlamini, F. M. Santucci, and P. Warren, eds.), pp. 1–22. FAO, Rome.

Sønstebø, J. H., Borgstrøm, R., and Heun, M. (2007). Genetic structure of brown trout (*Salmo trutta* L.) from the Hardangervidda mountain plateau (Norway) analyzed by microsatellite DNA: A basis for conservation guidelines. *Conserv. Genet.* **8,** 33–44.

Sork, V. L., and Smouse, P. E. (2006). Genetic analysis of landscape connectivity in tree populations. *Landsc. Ecol.* **21,** 821–836.

Sósa-Gomez, D. R., Moscardi, F., Santos, B., Alves, L. F. A., and Alves, S. B. (2008). Produção e uso de vírus para o controle de pragas na América Latina. *In* "Controle Microbiano de Pragas na América Latina" (S. B. Alves and R. B. Lopes, eds.), pp. 49–68. FEALQ, Piracicaba.

Souza, P. M., and Magalhães, P. O. (2010). Application of microbial α-amylase in industry—A review. *Braz. J. Microbiol.* **41,** 850–861.

Sparks, D., and Yates, I. E. (2002). Pecan pollen stored over a decade retains viability. *HortScience* **37,** 176–177.

Spooner, D., van Treuren, R., and de Vicente, M. C. (2005). Molecular markers for genebank management. *IPGRI Technical Bulletin No. 10*. International Plant Genetic Resources Institute, Rome.

Srivastava, P., and Chaturvedi, R. (2008). *In vitro* androgenesis in tree species: An update and prospect for further research. *Biotechnol. Adv.* **26,** 482–491.

Star, B., Nederbragt, A. J., Jentoft, S., Grimholt, U., Malmstrøm, M., Gregers, T. F., Rounge, T. B., Paulsen, J., Solbakken, M. H., Sharma, A., *et al.* (2011). The genome sequence of Atlantic cod reveals a unique immune system. *Nature* **477,** 207–210.

Stark, R., and Made, D. (2007). Detection of *Salmonella* in foods—Experience with a combination of microbiological and molecular biological methods. *Fleischwirtschaft* **87,** 98–101.

St-Cyr, J., Derome, N., and Bernatchez, L. (2008). The transcriptomics of life-history trade-offs in whitefish species pairs (*Coregonus* sp.). *Mol. Ecol.* **17,** 1850–1870.

Steane, D. A., Nicolle, D., Sansaloni, C. P., Petroli, C. D., Carling, J., Kilian, A., Myburg, A. A., Grattapaglia, D., and Vaillancourt, R. E. (2011). Population genetic analysis and phylogeny reconstruction in *Eucalyptus* (Myrtaceae) using high-throughput, genome-wide genotyping. *Mol. Phylogenet. Evol.* **59,** 206–224.

Steinkraus, K. H. (2002). Fermentations in world food processing. *Compr. Rev. Food Sci. Food Saf.* **1,** 23–32.

Stella, A., Ajmone-Marsan, P., Lazzari, B., and Boettcher, P. (2010). Identification of selection signatures in cattle breeds selected for dairy production. *Genetics* **185,** 1451–1461.

Stevanovic, J., Stanimirovic, Z., Dimitrijevic, V., and Maletic, M. (2010). Evaluation of 11 microsatellite loci for their use in paternity testing in Yugoslav Pied cattle (YU Simmental cattle). *Czech J. Anim. Sci.* **55,** 221–226.

Stothard, P., Choi, J. W., Basu, U., Sumner-Thomson, J. M., Meng, Y., Liao, X., and Moore, S. S. (2011). Whole genome resequencing of Black Angus and Holstein cattle for SNP and CNV discovery. *BMC Genomics* **12,** 559.

Stout, L., and Nüsslein, K. (2010). Biotechnological potential of aquatic plant-microbe interactions. *Curr. Opin. Biotechnol.* **21,** 339–345.

Streatfield, S. J., and Howard, J. A. (2003). Plant based vaccines. *Int. J. Parasitol.* **33,** 479–493.

Stroud, B. (2010). The year 2009 worldwide statistics of embryo transfer in domestic farm animals. Data Retrieval Committee Annual Report. *IETS Newsl.* **28,** 11–21.

Suen, G., Weimer, P. J., Stevenson, D. M., Aylward, F. O., Boyum, J., Deneke, J., Drinkwater, C., Ivanova, N. N., Mikhailova, N., Chertkov, O., *et al.* (2011). The complete genome sequence of *Fibrobacter succinogenes* S85 reveals a cellulolytic and metabolic specialist. *PLoS One* **6,** e18814.

Summer, L. W. (2010). Recent advances in plant metabolomics and greener pastures. *F1000 Biol. Rep.* **2,** 7.

Sun, Y.-Z., Yang, H.-L., Ma, R.-L., Zhang, C.-X., and Lin, W.-Y. (2011). Effect of dietary administration of Psychrobacter sp. on the growth, feed utilization, digestive enzymes and immune responses of grouper *Epinephelus coioides*. *Aquacult. Nutr.* **17,** e733–e740.

Swamy, B. P., and Sarla, N. (2008). Yield-enhancing quantitative trait loci (QTLs) from wild species. *Biotechnol. Adv.* **26,** 106–120.

Taborsky, M. (2001). The evolution of bourgeois, parasitic, and cooperative reproductive behaviors in fishes. *J. Hered.* **92,** 100–110.

Taggart, J. B., Bron, J. E., Martin, S. A. M., Seear, P. J., Høyheim, B., Talbot, R., Carmichael, S. N., Villeneuve, L., Sweeney, G. E., Houlihan, D. F., *et al.* (2008). A description of the origins, design and performance of the TRAITS–SGP Atlantic salmon Salmo salar L. cDNA microarray. *J. Fish Biol.* **72,** 2071–2094.

Takeya, M., Yamasaki, F., Uzuhashi, S., Aoki, T., Sawada, H., Nagai, T., Tomioka, K., Tomooka, N., Sato, T., and Kawase, M. (2011). NIASGBdb: NIAS Genebank databases for genetic resources and plant disease information. *Nucleic Acids Res.* **39,** D1108–D1113.

Tanaka, N., Abe, T., Miyazaki, S., and Sugawara, H. (2006). G-InforBIO: Integrated system for microbial genomics. *BMC Bioinformatics* **7,** 368.

Tanksley, S. D., and McCouch, S. R. (1997). Seed banks and molecular maps: Unlocking genetic potential from the wild. *Science* **277,** 1063–1066.

Tapio, M., Ozerov, M., Tapio, I., Toro, M. A., Marzanov, N., Ćinkulov, M., Goncharenko, G., Kiselyova, T., Murawski, M., and Kantanen, J. (2010). Microsatellite-based genetic diversity and population structure of domestic sheep in northern Eurasia. *BMC Genet.* **11,** 76–86.

Taras, D., Vahjen, W., Macha, M., and Simon, O. (2006). Performance, diarrhoea incidence, and occurrence of *Escherichia coli* virulence genes during long-term administration of a probiotic *Enterococcus faecium* strain to sows and piglets. *J. Anim. Sci.* **84,** 608–617.

Teletchea, F. (2009). Molecular identification methods of fish species: Reassessment and possible applications. *Rev. Fish Biol. Fish.* **19,** 265–293.

Teneva, A. (2009). Molecular markers in animal genome analysis. *Biotechnol. Anim. Husbandry* **25,** 1267–1284.

Tervit, H. R., Adams, S. L., Roberts, R. D., McGowan, L. T., Pugh, P. A., Smith, J. F., and Janke, A. R. (2005). Successful cryopreservation of Pacific oyster (*Crassostrea gigas*) oocytes. *Cryobiology* **51,** 142–151.

The Bovine Genome Sequencing and Analysis Consortium (2009). The genome sequence of Taurine cattle: A window to ruminant biology and evolution. *Science* **324,** 522–528.

The Bovine HapMap Consortium (2009). Genome-wide survey of SNP variation uncovers the genetic structure of cattle breeds. *Science* **324,** 528–532.

Thévenon, S., Dayo, G. K., Sylla, S., Sidibe, I., Berthier, D., Legros, H., Boichard, D., Eggen, A., and Gautier, M. (2007). The extent of linkage disequilibrium in a large cattle population of western Africa and its consequences for association studies. *Anim. Genet.* **38,** 277–286.

Thibier, M., and Wagner, H. G. (2002). World statistics for artificial insemination in cattle. *Livestock Prod. Sci.* **74,** 203–212.

Thirunavoukkarasu, M., Panda, P. K., Nayak, P., Behera, P. R., and Satpathy, G. B. (2010). Effect of media type and explant source on micropropagation of *Dalbergia sissoo* Roxb.—An important multipurpose forest tree. *Int. Res. J. Plant Sci.* **1,** 155–162.

Thomas, T. D., and Chaturvedi, R. (2008). Endosperm culture: A novel method for triploid plant production. *Plant Cell Tissue Organ Cult.* **93,** 1–14.

Thomas, T. D., Bhatnagar, A. K., and Bhojwani, S. S. (2000). Production of triploid plants of mulberry (*Morus alba* L.) by endosperm culture. *Plant Cell Rep.* **19,** 395–399.

Thomas, P. C., Divya, P. R., Chandrika, V., and Paulton, M. P. (2009). Genetic characterization of *Aeromonas hydrophila* using protein profiling and RAPD PCR. *Asian Fish. Sci.* **22,** 763–771.

Tian, F., Bradbury, P. J., Brown, P. J., Hung, H., Sun, Q., Flint-Garcia, S., Rocheford, T. R., McMullen, M. D., Holland, J. B., and Buckler, E. S. (2011). Genome-wide association study of leaf architecture in the maize nested association mapping population. *Nat. Genet.* **43,** 159–162.

Tiersch, T. R. (2008). Strategies for commercialization of cryopreserved fish semen. *R. Bras. Zootec.* **37,** 15–19.

Tiersch, T. R., Yang, H., Jenkins, J. A., and Dong, Q. (2007). Sperm cryopreservation in fish and shellfish. *Soc. Reprod. Fertil. Suppl.* **65,** 493–508.

Timmerman, H. M., Veldman, A., van den Elsen, E., Rombouts, F. M., and Beynen, A. C. (2006). Mortality and growth performance of broilers given drinking water supplemented with chicken-specific probiotics. *Poult. Sci.* **85,** 1383–1388.

Tinker, N. A., Kilian, A., Wight, C. P., Heller-Uszynska, K., Wenzl, P., Rines, H. W., Bjornstad, A., Howarth, C. J., Jannink, J. L., Anderson, J. M., *et al.* (2009). New DArT markers for oat provide enhanced map coverage and global germplasm characterization. *BMC Genomics* **10,** 39–60.

Tiquia, S. M. (2005). Microbial community dynamics in manure composts based on 16S and 18S rDNA T-RFLP profiles. *Environ. Technol.* **26,** 1101–1113.

Tixier-Boichard, M., Audiot, A., Bernigaud, R., Rognon, X., Berthouly, C., Magdelaine, P., Coquerelle, G., Grinand, R., Boulay, M., Ramanatseheno, D., *et al.* (2006). Valorisation des races anciennes de poulets: Facteurs sociaux, technicoéconomiques, génétiques et réglementaires. *Les Actes du BRG* **6,** 495–520.

Toosi, A., Fernando, R. L., and Dekkers, J. C. M. (2010). Genomic selection in admixed and crossbred populations. *J. Anim. Sci.* **88,** 32–46.

Toral Ibañez, M., Caru, M., Herrera, M. A., Gonzalez, L., Martin, L. M., Miranda, J., and Navarro-Cerrillo, R. M. (2009). Clones identification of *Sequoia sempervirens* (D. Don) Endl. in Chile by using PCR-RAPDs technique. *J. Zhejiang Univ. Sci. B* **10,** 112–119.

Toranzo, A. E., Romalde, J. L., Magariños, B., and Barja, J. L. (2009). Present and future of aquaculture vaccines against fish bacterial diseases. *Opt. Méditerranéennes*, A **86,** 155–176.

Torriani, S., Felis, G. E., and Fracchetti, F. (2011). Selection criteria and tools for malolactic starters development: An update. *Ann. Microbiol.* **61,** 33–39.

Touchell, D. H., and Dixon, K. W. (1994). Cryopreservation for seedbanking of Australian species. *Ann. Bot.* **40,** 541–546.

Tovar-Sánchez, E., and Oyama, K. (2004). Natural hybridization and hybrid zones between *Quercus crassifolia* and *Quercus crassipes* (Fagaceae) in Mexico: Morphological and molecular evidence. *Am. J. Bot.* **91,** 1352–1363.

Towill, L. E., and Walters, C. (2000). Cryopreservation of pollen. *In* "Cryopreservation of Tropical Plant Germplasm—Current Research Progress and Applications" (F. Engelmann and H. Takagi, eds.), pp. 115–129. JIRCAS, Tsukuba.

Troy, C. S., MacHugh, D. E., Bailey, J. F., Magee, D. A., Loftus, R. T., Cunningham, P., Chamberlain, A. T., Sykes, B. C., and Bradley, D. G. (2001). Genetic evidence for Near-Eastern origins of European cattle. *Nature* **410,** 1088–1091.

Tsai, C. J., and Hubscher, S. L. (2004). Cryopreservation in *Populus* functional genomics. *New Phytol.* **164,** 73–81.

Tsutsumi, H., Kono, M., Takai, K., Manabe, T., Haraguchi, M., Yamamoto, I., and Oppenheimer, C. (2000). Bioremediation on the shore after an oil spill from the Nakhodka in the Sea of Japan. III. Field tests of a bioremediation agent with microbiological cultures for the treatment of an oil spill. *Mar. Pollut. Bull.* **40,** 320–324.

Tuggle, C. K., Bearson, S. M., Uthe, J. J., Huang, T. H., Couture, O. P., Wang, Y. F., Kuhar, D., Lunney, J. K., and Honavar, V. (2010). Methods for transcriptomic analyses of the porcine host immune response: Application to Salmonella infection using microarrays. *Vet. Immunol. Immunopathol.* **138,** 280–291.

Tuskan, G. A., DiFazio, S., Jansson, S., Bohlmann, J., Grigoriev, I., Hellsten, U., Putnam, N., Ralph, S., Rombauts, S., Salamov, A., *et al.* (2006). The genome of black cottonwood, *Populus trichocarpa* (Torr. & Gray). *Science* **313,** 1596–1604.

Tweddle, J. C., Dickie, J. B., Baskin, C. C., and Baskin, J. M. (2003). Ecological aspects of seed desiccation sensitivity. *J. Ecol.* **91,** 294–304.

Tyagi, M., da Fonseca, M. M., and de Carvalho, C. C. (2011). Bioaugmentation and biostimulation strategies to improve the effectiveness of bioremediation processes. *Biodegradation* **22,** 231–241.

Tymchuk, W., O'Reilly, P. T., Bittman, J., MacDonald, D., and Schulte, P. (2010). Conservation genomics of Atlantic salmon: Variation in gene expression between and within regions of the Bay of Fundy. *Mol. Ecol.* **19,** 1842–1859.

Ueno, S., Le Provost, G., Léger, V., Klopp, C., Noirot, C., Frigerio, J. M., Salin, F., Salse, J., Abrouk, M., Murat, F., *et al.* (2010). Bioinformatic analysis of ESTs collected by Sanger and pyrosequencing methods for a keystone forest tree species: Oak. *BMC Genomics* **11,** 650.

Uimari, P., and Tapio, M. (2011). Extent of linkage disequilibrium and effective population size in Finnish Landrace and Finnish Yorkshire pig breeds. *J. Anim. Sci.* **89,** 609–614.

Ulrich, P., and Nowshari, M. (2002). Successful direct transfer of a deep frozen-thawed equine embryo. *Dtsch. Tierarztl. Wochenschr.* **109,** 61–62.

Umesh, N. R., Mohan, A. B. C., Ravibabu, G., Padiyar, P. A., Phillips, M. J., Mohan, C. V., and Bhat, B. V. (2010). Shrimp farmers in India: Empowering small-scale farmers through a cluster-based approach. *In* "Success Stories in Asian Aquaculture" (S. S. De Silva and F. B. Davy, eds.), pp. 41–66. Network of Aquaculture Centres in Asia Pacific, Bangkok.

Upadhyaya, H. D., Dwivedi, S. L., Baum, M., Varshney, R. K., Udupa, S. M., Gowda, C. L., Hoisington, D., and Singh, S. (2008). Genetic structure, diversity, and allelic richness in composite collection and reference set in chickpea (*Cicer arietinum* L.). *BMC Plant Biol.* **8,** 106.

Uyen, N. V., Ho, T. V., Tung, P. X., Vander Zaag, P., and Walker, T. S. (1996). Economic impact of the rapid multiplication of high-yielding, late-blight resistant varieties in Dalat, Vietnam. *In* "Case Studies of the Economic Impact of CIP-Related Technologies" (T. S. Walker and C. C. Crissman, eds.), pp. 127–138. International Potato Center, Lima.

Uyoh, E. A., Nkang, A. E., and Eneobong, E. E. (2003). Biotechnology, genetic conservation and sustainable use of bioresources. *Afr. J. Biotechnol.* **2,** 704–709.

Vaishampayan, A., Sinha, R. P., Hader, D.-P., Dey, T., Gupta, A. K., Bhan, U., and Rao, A. L. (2001). Cyanobacterial biofertilizers in rice agriculture. *Bot. Rev.* **67,** 453–516.

Vajta, G. (2000). Vitrification of the oocytes and embryos of domestic animals. *Anim. Reprod. Sci.* **60–61,** 357–364.

Van Damme, V., Gomez-Paniagua, H., and de Vicente, M. C. (2011). The GCP molecular marker toolkit, an instrument for use in breeding food security crops. *Mol. Breed.* **28,** 597–610.

Vandeputte, M., Kocour, M., Mauger, S., Dupont-Nivet, M., De Guerry, D., Rodina, M., Gela, D., Vallod, D., Chevassus, B., and Linhart, O. (2004). Heritability estimates for growth-related traits using microsatellite parentage assignment in juvenile common carp (*Cyprinus carpio* L.). *Aquaculture* **235,** 223–236.

Van Diepen, L. T. A., Lilleskov, E. A., and Pregitzer, K. S. (2011). Simulated nitrogen deposition affects community structure of arbuscular mycorrhizal fungi in northern hardwood forests. *Mol. Ecol.* **20,** 799–811.

Van Doornik, D. M., Waples, R. S., Baird, M. C., Moran, P., and Berntson, E. A. (2011). Genetic monitoring reveals genetic stability within and among threatened Chinook salmon populations in the Salmon River, Idaho. *N. Am. J. Fish. Manag.* **31,** 96–105.

Van Elsas, J. D., and Boersma, F. D. H. (2011). A review of molecular methods to study the microbiota of soil and the mycosphere. *Eur. J. Soil Biol.* **47,** 77–87.

Van Inghelandt, D., Melchinger, A. E., Lebreton, C., and Stich, B. (2010). Population structure and genetic diversity in a commercial maize breeding program assessed with SSR and SNP markers. *Theor. Appl. Genet.* **120,** 1289–1299.

Van Treuren, R., Magda, A., Hoekstra, R., and van Hintum, T. J. L. (2004). Genetic and economic aspects of marker-assisted reduction of redundancy from a wild potato germplasm collection. *Genet. Resour. Crop Evol.* **51,** 277–290.

Van Treuren, R., Boukema, I. W., de Groot, E. C., van de Wiel, C. C. M., and van Hintum, Th. J. L. (2010). Marker-assisted reduction of redundancy in a genebank collection of cultivated lettuce. *Plant Genet. Resour.* **8,** 95–105.

Vargas, A. M., Quesada Ocampo, L. M., Céspedes, M. C., Carreño, N., González, A., Rojas, A., Zuluaga, A. P., Myers, K., Fry, W. E., Jiménez, P., *et al.* (2009). Characterization of *Phytophthora infestans* populations in Colombia: First report of the A2 mating type. *Phytopathology* **99,** 82–88.

Vargas-Terán, M., Hofmann, H. C., and Tweddle, N. E. (2005). Impact of screwworm eradication programmes using the sterile insect technique. *In* "Sterile Insect Technique. Principles and Practice in Area-Wide Integrated Pest Management" (V. A. Dyck, J. Hendrichs, and A. S. Robinson, eds.), pp. 629–650. Springer, The Netherlands.

Varshney, R. K., Nayak, S. N., May, G. D., and Jackson, S. A. (2009a). Next-generation sequencing technologies and their implications for crop genetics and breeding. *Trends Biotechnol.* **27,** 522–530.

Varshney, R. K., Close, T. J., Singh, N. K., Hoisington, D. A., and Cook, D. R. (2009b). Orphan legume crops enter the genomics era!. *Curr. Opin. Plant Biol.* **12,** 1–9.

Varshney, R. K., Chen, W., Li, Y., Bharti, A. K., Saxena, R. K., Schlueter, J. A., Donoghue, M. T. A., Azam, S., Fan, G., Whaley, A. M., *et al.* (2012). Draft genome sequence of pigeonpea (Cajanus cajan), an orphan legume crop of resource-poor farmers. *Nat. Biotechnol.* **30,** 83–89. doi:10.1038/nbt.2022.

Vaseeharan, B., Prem Anand, T., Murugan, T., and Chen, J. C. (2006). Shrimp vaccination trials with the VP292 protein of white spot syndrome virus. *Lett. Appl. Microbiol.* **43,** 137–142.

Velusamy, V., Arshak, K., Korostynska, O., Oliwa, K., and Adley, C. (2010). An overview of foodborne pathogen detection: In the perspective of biosensors. *Biotechnol. Adv.* **28,** 232–254.

Vendrell, D., Balcazar, J. L., de Blas, I., Ruiz-Zarzuela, I., Girones, O., and Muzquiz, J. L. (2008). Protection of rainbow trout (*Oncorhynchus mykiss*) from lactococcosis by probiotic bacteria. *Comp. Immunol. Microbiol. Infect. Dis.* **31,** 337e45.

Vieitez, A. M., Corredoira, E., Ballester, A., Munoz, F., Duran, J., and Ibarra, M. (2009). *In vitro* regeneration of the important North American oak species *Quercus alba, Quercus bicolor and Quercus rubra. Plant Cell Tissue Organ Cult.* **98,** 135–145.

Vila, B., Fontgibell, A., Badiola, I., Esteve-Garcia, E., Jiménez, G., Castillo, M., and Brufau, J. (2009). Reduction of *Salmonella enterica* var. enteritidis colonization and invasion by *Bacillus cereus* var. toyoi inclusion in poultry feeds. *Poult. Sci.* **88,** 975–979.

Vila, B., Esteve-Garcia, E., and Brufau, J. (2010). Probiotic micro-organisms: 100 years of innovation and efficacy; modes of action. *Worlds Poult. Sci. J.* **65,** 369–380.

Vincelli, P., and Tisserat, N. (2008). Nucleic acid-based pathogen detection in applied plant pathology. *Plant Dis.* **92,** 660–669.

Vine, N. G., Leukes, W. D., and Kaiser, H. (2006). Probiotics in marine larviculture. *FEMS Microbiol. Rev.* **30,** 404–427.

Vingborg, R. K. K., Gregersen, V. R., Zhan, B., Panitz, F., Høj, A., Sørensen, K. K., Madsen, L. B., Larsen, K., Hornshøj, H., Wang, X., *et al.* (2009). A robust linkage map of the porcine autosomes based on gene-associated SNPs. *BMC Genomics* **10,** 134.

Visser, B., Herselman, L., and Pretorius, Z. A. (2009). Genetic comparison of Ug99 with selected South African races of Puccinia graminis f.sp. tritici. *Mol. Plant Pathol.* **10,** 213–222.

Visser, B., Herselman, L., Park, R. F., Karaoglu, H., Bender, C. M., and Pretorius, Z. A. (2010). Characterization of two new wheat stem rust races within the Ug99 lineage in South Africa. BGRI 2010 Technical Workshop, 30-31- May 2010, St Petersburg, Russia.

Viveiros, A. T. M., and Godinho, H. P. (2009). Sperm quality and cryopreservation of Brazilian freshwater fish species: A review. *Fish Phys. Biochem.* **35,** 137–150.

Vizcaíno, J. A., Foster, J. M., and Martens, L. (2010). Proteomics data repositories: Providing a safe haven for your data and acting as a springboard for further research. *J. Proteomics* **73,** 2136–2146.

Vogel, B. (2009). Marker-assisted selection: A non-invasive biotechnology alternative to genetic engineering of plant varieties. *Report prepared for Greenpeace International.* http://www.greenpeace.org/raw/content/international/press/reports/smart-breeding.pdf.

Volis, S., and Blecher, M. (2010). Quasi *in situ*: A bridge between *ex situ* and *in situ* conservation of plants. *Biodivers. Conserv.* **19,** 2441–2454.

Volk, G. M. (2010). Application of functional genomics and proteomics to plant cryopreservation. *Curr. Genomics* **11,** 24–29.

Volkaert, H., Lowe, A., Davies, S., Cavers, S., Indira, E. P., Sudarsono, S., Vanavichit, A., Van Der Straeten, D., and Wellendorf, H. (2008). Developing know-how for the improvement and sustainable management of teak genetic resources. http://nora.nerc.ac.uk/7688/.

Von Buenau, R., Jaekel, L., Schubotz, E., Schwarz, S., Stroff, T., and Krueger, M. (2005). *Escherichia coli* strain Nissle 1917: Significant reduction of neonatal calf diarrhoea. *J. Dairy Sci.* **88,** 317–323.

Von Felten, A., Meyer, J. B., Défago, G., and Maurhofer, M. (2011). Novel T-RFLP method to investigate six main groups of 2,4-diacetylphloroglucinol-producing pseudomonads in environmental samples. *J. Microbiol. Methods* **84,** 379–387.

Votava, E. J., Nabhan, G. P., and Bosland, P. W. (2002). Genetic diversity and similarity revealed via molecular analysis among and within an *in situ* population and *ex situ* accessions of chiltepin (*Capsicum annuum* var. glabriusculum). *Conserv. Genet.* **3,** 123–129.

Vreysen, M. J. B., and Robinson, A. S. (2010). Ionizing radiation and area-wide management of insect pests to promote sustainable agriculture: A review. *Agron. Sustain. Dev.* **31,** 233–250. doi:10.1051/agro/20100009.

Vrijenhoek, R. C. (2009). Cryptic species, phenotypic plasticity, and complex life histories: Assessing deep-sea faunal diversity with molecular markers. *Deep Sea Res. Part 2 Top. Stud. Oceanogr.* **56,** 1713–1723.

Vurro, M., Bonciani, B., and Vannacci, G. (2010). Emerging infectious diseases of crop plants in developing countries: Impact on agriculture and socio-economic consequences. *Food Secur.* **2,** 113–132.

Wade, C. M., Giulotto, E., Sigurdsson, S., Zoli, M., Gnerre, S., Imsland, F., Lear, T. L., Adelson, D. L., Bailey, E., Bellone, R. R., *et al.* (2009). Genome sequence, comparative analysis, and population genetics of the domestic horse. *Science* **326,** 865–867.

Wahab, S. (2009). Biotechnological approaches in the management of plant pests, diseases and weeds for Sustainable Agriculture. *J. Biopesticides* **22,** 115–134.

Walters, C., Wheeler, L. J., and Stanwood, P. C. (2004). Longevity of cryogenically-stored seeds. *Cryobiology* **48,** 229–244.

Wambugu, F. (2004). Food, nutrition and economic empowerment: the case for scaling up the tissue culture banana project to the rest of Africa. Paper presented at the NEPAD/IGAD regional conference agricultural successes in the Greater Horn of Africa. pp. 22–25. Nairobi, Kenya. November 2004.

Wang, J. (2005). Estimation of effective population sizes from data on genetic markers. *Philos. Trans. R. Soc. Lond. B Biol. Sci.* **360,** 1395–1409.

Wang, X. R., and Szmidt, A. E. (2001). Molecular markers in population genetics of forest trees. *Scand. J. For. Res.* **16,** 199–220.

Wang, Q. C., Panis, B., Engelmann, F., Lambardi, M., and Valkonen, J. P. T. (2009a). Cryotherapy of shoot tips: A technique for pathogen eradication to produce healthy planting materials and prepare healthy plant genetic resources for cryopreservation. *Ann. Appl. Biol.* **154,** 351–363.

Wang, M. J., Qi, X. L., Zhao, S. T., Zhang, S. G., and Lu, M. Z. (2009b). Dynamic changes in transcripts during regeneration of the secondary vascular system in *Populus tomentosa* Carr. revealed by cDNA microarrays. *BMC Genomics* **10,** 215.

Wang, J. M., Yang, J. M., Zhu, J. H., Jia, Q. J., and Tao, Y. Z. (2010a). Assessment of genetic diversity by simple sequence repeat markers among forty elite varieties in the germplasm for malting barley breeding. *J. Zhejiang Univ. Sci. B* **1,** 792–800.

Wang, X., Torimaru, T., Lindgren, D., and Fries, A. (2010b). Marker-based parentage analysis facilitates low input 'breeding without breeding' strategies for forest trees. *Tree Genet. Genomes* **6,** 227–235.

Wang, H., Li, X., Wang, M., Clarke, S., Gluis, M., and Zhang, Z. (2011). Effects of larval cryopreservation on subsequent development of the blue mussels, *Mytilus galloprovincialis* Lamarck. *Aquacult. Res.* **42,** 1816–1823.

Wannaprasat, W., Koowatananukul, C., Ekkapobyotin, C., and Chuanchuen, R. (2009). Quality analysis of commercial probiotic products for food animals. *Southeast Asian J. Trop. Med. Public Health* **40,** 1103–1112.

Ward, R. D., Hanner, R., and Hebert, P. D. N. (2009). The campaign to DNA barcode all fishes, FISH-BOL. *J. Fish Biol.* **74,** 329–356.

Watt, M. P., Thokoane, N. L., Mycock, D., and Blakeway, F. (2000). *In vitro* storage of *Eucalyptus grandis* germplasm under minimal growth conditions. *Plant Cell Tissue Organ Cult.* **61,** 161–164.

Wegrzyn, J. L., Lee, J. M., Tearse, B. R., and Neale, D. B. (2008). TreeGenes: A forest tree genome database. *Int. J. Plant Genomics* **2008,** 412875.

Wei, X., Yuan, X., Yu, H., Wang, Y., Xu, Q., and Tang, S. (2009). Temporal changes in SSR allelic diversity of major rice cultivars in China. *J. Genet. Genomics* **36,** 363–370.

Weichselbaum, E. (2009). Probiotics and health: A review of the evidence. *Nutr. Bull.* **34,** 340–373.

Weil, C. F. (2009). TILLING in grass species. *Plant Physiol.* **149,** 158–164.

Wenne, R., Boudry, P., Hemmer-Hansen, J., Lubieniecki, K. P., Was, A., and Kause, A. (2007). What role for genomics in fisheries management and aquaculture? *Aquat. Living Resour.* **20,** 241–255.

Wennevik, V., Jørstad, K. E., Dahle, G., and Fevolden, S. E. (2008). Mixed stock analysis and the power of different classes of molecular markers in discriminating coastal and oceanic Atlantic cod (*Gadus morhua* L.) on the Lofoten spawning grounds, Northern Norway. *Hydrobiologia* **606,** 7–25.

Wenzel, W. (2009). Rhizosphere processes and management in plant-assisted bioremediation (phytoremediation) of soils. *Plant Soil* **321,** 385–408.

Wenzl, P., Carling, J., Kudrna, D., Jaccoud, D., Huttner, E., Kleinhofs, A., and Kilian, A. (2004). Diversity Arrays Technology (DArT) for whole-genome profiling of barley. *Proc. Natl. Acad. Sci. U.S.A.* **101,** 9915–9920.

Werner, C., Poontawee, K., Mueller-Belecke, A., Hoerstgen-Schwark, G., and Wicke, M. (2008). Flesh characteristics of pan-size triploid and diploid rainbow trout (*Oncorhynchus mykiss*) reared in a commercial fish farm. *Arch. Tierz. Arch. Anim. Breed.* **51,** 71–83.

Weyens, N., van der Lelie, D., Taghavi, S., and Vangronsveld, J. (2009). Phytoremediation: Plant-endophyte partnerships take the challenge. *Curr. Opin. Biotechnol.* **20,** 248–254.

WFCC (2010). World Federation for Culture Collections: Guidelines for the Establishment and Operation of Collections of Cultures of Microorganisms. 3rd edn. http://www.wfcc.info/guidelines/.

White, G. M., Boshier, D. H., and Powell, W. (2002). Increased pollen flow counteracts fragmentation in a tropical dry forest: An example from *Swietenia humilis* Zuccarini. *Proc. Natl. Acad. Sci. U.S.A.* **99,** 2038–2042.

Whiteley, A. R., Derome, N., Rogers, S. M., St-Cyr, J., Laroche, J., Labbe, A., Nolte, A., Renaut, S., Jeukens, J., and Bernatchez, L. (2008). The phenomics and expression quantitative trait locus mapping of brain transcriptomes regulating adaptive divergence in lake whitefish species pairs (*Coregonus* sp.). *Genetics* **180,** 147–164.

Wieland, G., Neumann, R., and Backhaus, H. (2001). Variation of microbial communities in soil, rhizosphere, and rhizoplane in response to crop species, soil type, and crop development. *Appl. Environ. Microbiol.* **67,** 5849–5854.

Wilhelm, V., Miquel, A., Burzio, L. O., Rosemblatt, M., Engel, E., Valenzuela, S., Parada, G., and Valenzuela, P. D. (2006). A vaccine against the salmonid pathogen *Piscirickettsia salmonis* based on recombinant proteins. *Vaccine* **24,** 5083–5091.

Wilkins, N. P., Cotter, D., and O'Maoiléidigh, N. (2001). Ocean migration and recaptures of tagged, triploid, mixed-sex and all-female Atlantic salmon (*Salmo salar* L.) released from rivers in Ireland. *Genetica* **111,** 197–212.

William, M., Morris, M., Warburton, M., and Hoisington, D. (2007). Technical, economic and policy considerations on marker-assisted selection in crops: Lessons from the experience at an international agricultural research center. *In* "Marker-assisted Selection: Current Status and Future Perspectives in Crops, Livestock, Forestry and Fish" (E. Guimarães, J. Ruane, B. Scherf, A. Sonnino, and J. Dargie, eds.), pp. 381–404. FAO, Rome.

Wilson, S. (2010). "Plant Doctors" a global prescription for plant pests. *Issues* **90,** 34–37.

Wilson, A. J., and Ferguson, M. M. (2002). Molecular pedigree analysis in natural populations of fishes: Approaches, applications, and practical considerations. *Can. J. Fish. Aquat. Sci.* **59,** 1696–1707.

Wimmers, K., Murani, E., and Ponsuksili, S. (2010). Functional genomics and genetical genomics approaches towards elucidating networks of genes affecting meat performance in pigs. *Brief. Funct. Genomics* **9,** 251–258.

Winans, G. A., Paquin, M. M., and van Doornik, D. M. (2004). Genetic stock identification of steelhead in the Columbia River Basin: An evaluation of different molecular markers. *N. Am. J. Fish. Manag.* **24,** 672–685.

Winter, J. M., Behnken, S., and Hertweck, C. (2011). Genomics-inspired discovery of natural products. *Curr. Opin. Chem. Biol.* **15,** 22–31.

Witteveldt, J., Vlak, J. M., and van Hulten, M. C. (2004). Protection of *Penaeus monodon* against white spot syndrome virus using a WSSV subunit vaccine. *Fish Shellfish Immunol.* **16,** 571–579.

Witteveldt, J., Jolink, M., Cifuentes, C. E., Vlak, J. M., and van Hulten, M. C. (2005). Vaccination of *Penaeus monodon* against white spot syndrome virus using structural virion proteins. *In* "Diseases in Asian Aquaculture V" (P. J. Walker, R. G. Lester, and M. G. Bondad-Reantaso, eds.), pp. 513–522. Asian Fisheries Society, Manila.

Wohlgemuth, S., Loh, G., and Blaut, M. (2010). Recent developments and perspectives in the investigation of probiotic effects. *Int. J. Med. Microbiol.* **300,** 3–10.

Wong, L. L., Peatman, E., Lu, J., Kucuktas, H., He, S., Zhou, C., Na-nakorn, U., and Liu, Z. (2011). DNA barcoding of catfish: Species authentication and phylogenetic assessment. *PLoS One* **6,** e17812.

Workenhe, S. T., Hori, T. S., Rise, M. L., Kibenge, M. J., and Kibenge, F. S. (2009). Infectious salmon anaemia virus (ISAV) isolates induce distinct gene expression responses in the Atlantic salmon (*Salmo salar*) macrophage/dendritic-like cell line TO, assessed using genomic techniques. *Mol. Immunol.* **46,** 2955–2974.

Wright, S. I., Bi, I. V., Schroeder, S. G., Yamasaki, M., Doebley, J. F., McMullen, M. D., and Gaut, B. S. (2005). The effects of artificial selection on the maize genome. *Science* **308,** 1310–1314.

Wu, M., Song, L., Ren, J., Kan, J., and Qian, P.-Y. (2004). Assessment of microbial dynamics in the Pearl River Estuary by 16S rRNA terminal restriction fragment analysis. *Cont. Shelf Res.* **24,** 1925–1934.

Wu, W. Z., Wang, X. Q., Wu, G. Y., Kim, S. W., Chen, F., and Wang, J. J. (2010). Differential composition of proteomes in sow colostrum and milk from anterior and posterior mammary glands. *J. Anim. Sci.* **88,** 2657–2664.

Wu, L. J., Wang, H. Q., Wang, E. T., Chen, W. X., and Tian, C. F. (2011). Genetic diversity of nodulating and non-nodulating rhizobia associated with wild soybean (*Glycine soja* Sieb. & Zucc.) in different ecoregions of China. *FEMS Microbiol. Ecol.* **76,** 439–450.

Xia, L., Peng, K., Yang, S., Wenzl, P., de Vicente, C. M., Fregene, M., and Kilian, A. (2005). DArT for high-throughput genotyping of cassava (*Manihot esculenta*) and its wild relatives. *Theor. Appl. Genet.* **110,** 1092–1098.

Xiang, L., He, D., Dong, W., Zhang, Y., and Shao, J. (2010). Deep sequencing-based transcriptome profiling analysis of bacteria-challenged *Lateolabrax japonicus* reveals insight into the immune-relevant genes in marine fish. *BMC Genomics* **11,** 472.

Xie, Y., McNally, K., Li, C., Leung, H., and Zhu, Y. (2006). A high-throughput genomic tool: Diversity array technology complementary for rice genotyping. *J. Integr. Plant Biol.* **48,** 1069–1076.

Xu, S., Tauer, C. G., and Nelson, C. D. (2008). Natural hybridization within seed sources of shortleaf pine (*Pinus echinata* Mill.) and loblolly pine (*Pinus taeda* L.). *Tree Genet. Genomes* **4,** 849–858.

Xu, L., Liu, C., Zhang, L., Wang, Z., Han, X., Li, X., and Chang, S. (2010). Genetic diversity in goat breeds based on microsatellite analysis. *Chin. J. Biotechnol.* **26,** 588–594.

Xu, F., Wang, J., Chen, S., Qin, W., Yu, Z., Zhao, H., Xing, X., and Li, H. (2011). Strain improvement for enhanced production of cellulase in *Trichoderma viride*. *Appl. Biochem. Microbiol.* **4,** 53–58.

Yadav, R. S., Sehgal, D., and Vadez, V. (2011a). Using genetic mapping and genomics approaches in understanding and improving drought tolerance in pearl millet. *J. Exp. Bot.* **62,** 397–408.

Yadav, J., Yadav, S., and Singh, S. G. (2011b). Plant growth promotion in wheat crop under environmental condition by PSB as bio-fertilizer. *Res. J. Agric. Sci.* **2,** 76–78.

Yamada, T., and Sekiguchi, Y. (2009). Cultivation of uncultured chloroflexi subphyla: Significance and ecophysiology of formerly uncultured chloroflexi 'subphylum i' with natural and biotechnological relevance. *Microbes Environ.* **24,** 205–216.

Yang, H., and Tiersch, T. R. (2009). Current status of sperm cryopreservation in biomedical research fish models: Zebrafish, medaka, and *Xiphophorus*. *Comp. Biochem. Physiol. C Toxicol. Pharmacol.* **149,** 224–232.

Yang, S., Pang, W., Ash, G., Harper, J., Carling, J., Wenzl, P., Huttner, E., and Kilian, A. (2006). Low level of genetic diversity in cultivated pigeonpea compared to its wild relatives is revealed by Diversity Arrays Technology (DArT). *Theor. Appl. Genet.* **113,** 585–595.

Yang, X., Kalluri, U. C., DiFazio, S. P., Wullschleger, S. D., Tschaplinski, T. J., Cheng, M. Z., and Tuskan, G. A. (2009a). Poplar genomics: State of the science. *CRC Crit. Rev. Plant Sci.* **28,** 285–308.

Yang, S., Tschaplinski, T. J., Engle, N. L., Carroll, S. L., Martin, S. L., Davison, B. H., Palumbo, A. V., Rodriguez, M., Jr., and Brown, S. D. (2009b). Transcriptomic and metabolomic profiling of *Zymomonas mobilis* during aerobic and anaerobic fermentations. *BMC Genomics* **10,** 34.

Yang, J., Kloepper, J. W., and Ryu, C. M. (2009c). Rhizosphere bacteria help plants tolerate abiotic stress. *Trends Plant Sci.* **14,** 1–4.

Yasodha, R., Sumathi, R., and Gurumurthi, K. (2004). Micropropagation for quality propagule production in plantation forestry. *Indian J. Biotechnol.* **3,** 159–170.

Yasui, G. S., Fujimotoa, T., and Araia, K. (2010). Restoration of the loach, *Misgurnus anguillicaudatus*, from cryopreserved diploid sperm and induced androgenesis. *Aquaculture* **308**(Suppl. 1), S140–S144.

Yoshikawa, H., Morishima, K., Fujimoto, T., Arias-Rodriguez, L., Yamaha, E., and Arai, K. (2008). Ploidy manipulation using diploid sperm in the loach, Misgurnus anguillicaudatus: A review. *J. Appl. Ichthyol.* **24,** 410–414.

Yoshizaki, G., Fujinuma, K., Iwasaki, Y., Okutsu, T., Shikina, S., Yazawa, R., and Takeuchi, Y. (2011). Spermatogonial transplantation in fish: A novel method for the preservation of genetic resources. *Comp. Biochem. Physiol. Part D Genomics Proteomics* **6,** 55–61.

Yue, G. H., and Chang, A. (2010). Molecular evidence for high frequency of multiple paternity in a freshwater shrimp species *Caridina ensifera*. *PLoS One* **5,** e12721.

Yue, G. H., Li, J. L., Wang, C. M., Xia, J. H., Wang, G. L., and Feng, J. B. (2010). High prevalence of multiple paternity in the invasive crayfish species, *Procambarus clarkii*. *Int. J. Biol. Sci.* **6,** 107–115.

Yusibov, V., Streatfield, S. J., and Kushnir, N. (2011). Clinical development of plant-produced recombinant pharmaceuticals: Vaccines, antibodies and beyond. *Hum. Vaccin.* **7,** 1–9.

Zamir, D. (2001). Improving plant breeding with exotic genetic libraries. *Nat. Rev. Genet.* **2,** 983–989.

Zarghami, R., Pirseyedi, M., Hasrak, S., and Sardrood, B. P. (2008). Evaluation of genetic stability in cryopreserved *Solanum tuberosum*. *Afr. J. Biotechnol.* **7,** 2798–2802.

Zarkti, H., Ouabbou, H., Hilali, A., and Udupa, S. M. (2010). Detection of genetic diversity in Moroccan durum wheat accessions using agro-morphological traits and microsatellite markers. *Afr. J. Agric. Res.* **5,** 1837–1844.

Zeng, Y. F., Liao, W. J., Petit, R. J., and Zhang, D. Y. (2010). Exploring species limits in two closely related Chinese oaks. *PLoS One* **5,** e15529.

Zhang, D., Zhang, H., Yang, L., Guo, J., Li, X., and Feng, Y. (2009). Simultaneous detection of *Listeria monocytogenes*, *Staphylococcus aureus*, *Salmonella enterica* and *Escherichia coli* O157:H7 in food samples using multiplex PCR method. *J. Food Saf.* **29,** 348–363.

Zhang, W., Li, F., and Nie, L. (2010a). Integrating multiple 'omics' analysis for microbial biology: Application and methodologies. *Microbiology* **156,** 287–301.

Zhang, L., Xu, J. Q., Liu, H.-L., Lai, T., Ma, J.-L., Wang, J.-F., and Zhu, Y.-H. (2010b). Evaluation of *Lactobacillus rhamnosus* GG using an *Escherichia coli* K88 model of piglet diarrhoea: Effects on diarrhoea incidence, faecal microflora and immune responses. *Vet. Microbiol.* **141,** 142–148.

Zhang, H., Liu, S. J., Zhang, C., Tao, M., Peng, L. Y., You, C. P., Xiao, J., Zhou, Y., Zhou, G. J., Luo, K. K., *et al.* (2011a). Induced gynogenesis in grass carp (*Ctenopharyngodon idellus*) using irradiated sperm of allotetraploid hybrids. *Mar. Biotechnol.* **13,** 1017–1026.

Zhang, W., Dudley, E. G., and Wade, J. T. (2011b). Genomic and transcriptomic analyses of foodborne bacterial pathogens. *In* "Genomics of Foodborne Bacterial Pathogens, Food Microbiology and Food Safety" (M. Weidmann and W. Zhang, eds.), pp. 311–341. Springer, New York.

Zhong, S., Dekkers, J. C., Fernando, R. L., and Jannink, J. L. (2009). Factors affecting accuracy from genomic selection in populations derived from multiple inbred lines: A Barley case study. *Genetics* **182,** 355–364.

Zhou, X., Abbas, K., Li, M., Fang, L., Li, S., and Wang, W. (2010a). Comparative studies on survival and growth performance among diploid, triploid and tetraploid dojo loach *Misgurnus anguillicaudatus*. *Aquaculture Int.* **18,** 349–359.

Zhou, X., Tian, Z., Wang, Y., and Li, W. (2010b). Effect of treatment with probiotics as water additives on tilapia (*Oreochromis niloticus*) growth performance and immune response. *Fish Physiol. Biochem.* **36,** 501–509.

Zohar, Y., and Mylonas, C. C. (2001). Endocrine manipulations of spawning in cultured fish: From hormones to genes. *Aquaculture* **197,** 99–136.

Zordan, M. D., Grafton, M. M. G., and Leary, J. F. (2011). An integrated microfluidic biosensor for the rapid screening of foodborne pathogens by surface plasmon resonance imaging. *Progress in Biomedical Optics and Imaging—Proceedings of SPIE* **7888,** 78880R–78880R-10.

Zunabovic, M., Domig, K. J., and Kneifel, W. (2011). Practical relevance of methodologies for detecting and tracing of *Listeria monocytogenes* in ready-to-eat foods and manufacture environments—A review. *Food Sci. Technol.* **44,** 351–362.

2

Transposable Elements and Insecticide Resistance

Wayne G. Rostant, Nina Wedell, and David J. Hosken
Centre for Ecology and Conservation, University of Exeter, Cornwall Campus, Tremough, Penryn, Cornwall, United Kingdom

Advances in Genetics, Vol. 78
0065-2660/12 $35.00
http://dx.doi.org/10.1016/B978-0-12-394394-1.00002-X

ABSTRACT

Transposable elements (TEs) are mobile DNA sequences that are able to copy themselves within a host genome. They were initially characterized as selfish genes because of documented or presumed costs to host fitness, but it has become increasingly clear that not all TEs reduce host fitness. A good example of TEs benefiting hosts is seen with insecticide resistance, where in a number of cases, TE insertions near specific genes confer resistance to these man-made products. This is particularly true of *Accord* and associated TEs in *Drosophila melanogaster* and *Doc* insertions in *Drosophila simulans*. The first of these insertions also has sexually antagonistic fitness effects in the absence of insecticides, and although the magnitude of this effect depends on the genetic background in which *Accord* finds itself, this represents an excellent example of intralocus sexual conflict where the precise allele involved is well characterized. We discuss this finding and the role of TEs in insecticide resistance. We also highlight areas for further research, including the need for surveys of the prevalence and fitness consequences of the *Doc* insertion and how *Drosophila* can be used as models to investigate resistance in pest species.

I. INTRODUCTION

The concept of an essentially stable genome, with each specific genetic element confined to a single locus was developed during the first few decades of last century. This simple picture first came under serious challenge through the work of McClintock (1950, 1984) who, while analyzing chromosome breakage in maize at Cornell University, first discovered what we now know as transposable elements (TEs). McClintock called these mobile elements "controlling elements," a term which reveals her early assertion of their potential involvement in gene expression.

This view has turned out to be remarkably prescient. However, TEs have spent much of the time since their discovery under the monikers "junk DNA" and "selfish DNA," revealing a general opinion that these mobile stretches of DNA played little if any part in evolution of their "hosts." TEs were largely thought to have no influence on host genes and were interesting only insofar as their unique form of drive allowed them to invade host genomes and spread through populations. A recent review by Biémont (2010) gives a good account of how prevailing views have come full circle to vindicate McClintock's proposal that TEs are crucial components of genomes and drivers of their evolution through their ability to affect gene expression. This journey from junk to critical agents of adaptive change has gathered pace as the sequencing of whole genomes has revealed the ubiquity and diversity of TEs.

There was an initial reluctance by many geneticists to accept that maize was not an anomalous case. That so-called jumping genes might exist in other genomes was difficult to reconcile with the fact that genetic maps had revealed remarkable homogeneity between individuals within species. The success of mapping of genes to precise positions on a chromosome was incompatible with genes moving around the genome. The isolation of bacterial TEs from *Escherichia coli* (Shapiro, 1969) was the first step toward acceptance that TEs were a general feature of genomes.

In spite of these discoveries, it was not until the 1970s that skepticism over the fundamental importance of TEs finally began to erode (Biémont and Vieira, 2006). The reason for this was the emergence of hybrid dysgenesis—a phenomenon observed when females of laboratory *Drosophila melanogaster* stocks were mated with males derived from natural populations. The progeny of these crosses displayed unusual germ line phenotypes including sterility, high mutation rate, and increased frequency of chromosomal aberration, while no such deficiencies exist in the reciprocal cross. The source of the dysgenesis turned out to be a TE called the *P* element which was present in wild strains but absent in laboratory strains.

Concurrent with a developing understanding of the ubiquity and importance of TEs within genomes has been the increased use of pesticides to control pest organisms, particularly from the 1950s onward (Wilson, 2001). This strong pervasive selection over many generations has provided the theoretical conditions under which adaptation by major genes might be favored, although early models suggested that strong selection, while necessary, is probably insufficient to favor major gene over polygenic adaptation (Lande, 1983). Nevertheless, the overwhelming empirical evidence is that the evolution of pesticide resistance is most often associated with the spread of a major mutation (Wilson, 2001), and it has been suggested that it is not the strength of selection *per se*, but the amount of phenotypic change required to achieve adaptation which determines the genetic architecture of the adaptive response (Macnair, 1991).

In light of these theoretical and empirical findings, it is perhaps unsurprising that TEs are increasingly being implicated in the adaptive response of organisms to man-made xenobiotics. In this review, we highlight the properties of TEs and insecticide resistance that make the former uniquely suited to the latter adaptive response. While drawing on several putative and several well-documented examples of TE-mediated insecticide resistance from the literature, we focus primarily on the striking cases of DDT resistance in *D. melanogaster* and *Drosophila simulans* which have been particularly well studied. In both instances, the resistance phenotype has been conferred by parallel insertions of TEs near a cytochrome P450 gene.

II. TRANSPOSABLE ELEMENTS

A. Definition and origin

TEs, simply put, are DNA sequences that have the capacity to transpose. That is, they change their chromosomal location from one position to another within the same genome, within a single cell (Hua-Van *et al.*, 2011; Kidwell and Lisch, 2001). They typically encode genes to promote this movement, in which case transpositional ability is intrinsic. These TEs are said to be autonomous and contrast with nonautonomous TEs which cannot transpose on their own, instead depending on the transposition machinery of other TEs (Hua-Van *et al.*, 2011; Kidwell and Lisch, 2001; Wicker *et al.*, 2007).

While questions concerning the origin and early evolution of TEs may never be fully resolved, it does appear that their evolution has occurred primarily through the serial addition of domains, several of which seem likely to have evolved from bacteria (Kidwell and Lisch, 2001). The question of a common origin for all TEs remains open (Wicker *et al.*, 2007).

B. Classification of TEs

The first TE classification system was proposed by Finnegan (1989) and included two main TE classes which were distinguished by their transposition intermediate. Class I elements include those which transcribe via an RNA intermediate and, using a “copy-and-paste” mechanism, establish new copies of themselves elsewhere in a genome. Class II elements, in contrast, excise from donor sites and move to new locations in a genome without use of an RNA intermediate, that is, they use a “cut-and-paste” method of transposition (Fig. 2.1).

To cope with an expanded array of TEs with diverse characteristics, Wicker *et al.* (2007) proposed a classification scheme that built on Finnegan’s original proposal (Table 2.1) by incorporating mechanistic and enzymatic criteria to the classification procedure. The original two classes were retained, and two subclasses within class II (DNA transposons) were formed to separate DNA transposons which leave the donor site (excision) to reintegrate elsewhere (subclass 1) from those which copy themselves for insertion (subclass 2). The next hierarchical ranking (i.e., order) marks differences in the insertion mechanism and thus organization and enzymology. Superfamilies within an order

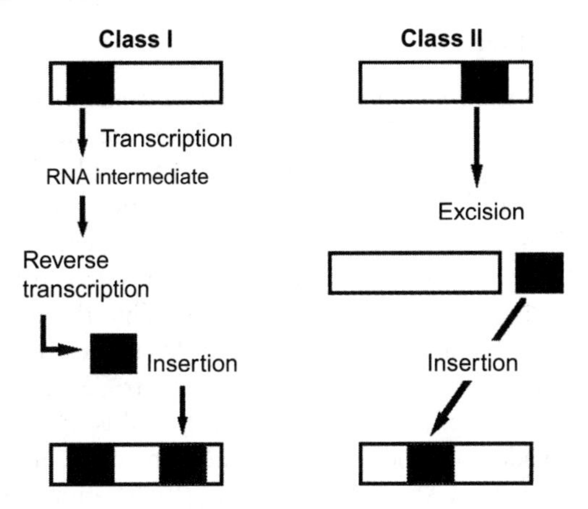

Figure 2.1. A diagram of the transpositional modes of the two major transposable element classes, class I and class II. Class I elements do not move once inserted into host DNA but use RNA intermediates to insert additional TE copies in new genomic locations. Class II elements can move by excision from host DNA, followed by reinsertion into a new location.

Table 2.1. The Transposable Element Classification System Proposed by Wicker *et al.* (2007)

Class	Order	Superfamilies
Class I (retrotransposons)	LTR (long-terminal repeat)	*Copia*, *Gypsy*, *Bel-Pao*, *Retrovirus*, *ERV*
	DIRS (*Dictyostelium* intermediate repeat sequence)	*DIRS*, *Ngaro*, *VIPER*
	PLE (*Penelope*-like elements)	*Penelope*
	LINE (long interspersed nuclear element)	*R2*, *RTE*, *Jockey*, *L1*, *I element*
	SINE (short interspersed nuclear element)	*tRNA*, *7SL*, *5S*
Class II (DNA transposons): subclass 1	TIR (terminal inverted repeat)	*Tc1-Mariner*, *hAT*, *Mutator*, *Merlin*, *Transib*, *P element*, *PiggyBac*, *PIF-Harbinger*, CACTA
	Crypton	*Crypton*
Class II (DNA transposons): subclass 2	Helitron	*Helitron*
	Maverick	*Maverick*

share a replication strategy but are distinguished by large-scale features such as the structure of protein or noncoding regions. Families within superfamilies are defined by DNA sequence conservation.

C. Transposition rates

The best data on TE transposition rates have come from laboratory experiments on *D. melanogaster* (Burt and Trivers, 2006). Rates are variable, ranging from 2.9×10^{-6} per element per generation for *P* elements in inbred lines (Dominguez and Albornoz, 1996) to 0.25 per *P* element in dysgenic crosses. However, they are typically low—estimates of 10^{-4} with the order of magnitude variation have been found for LINEs and LTR retrotransposons in two separate experiments (Maside *et al.*, 2000; Nuzhdin and Mackay, 1995). That these estimates tend to be higher than excision rates (of the order 10^{-6}) implies that TEs should, in general, accumulate in genomes over evolutionarily trivial timescales (Burt and Trivers, 2006). As a recent, well-cited example of this, *P* elements had been shown to have invaded all known wild populations of *D. melanogaster* in the matter of about 50 years (Anxolabéhère *et al.*, 1988) after horizontal transfer from *Drosophila willistoni* (Daniels *et al.*, 1990).

D. The abundance and distribution of TEs

TEs have been discovered and characterized in most species that have been adequately examined (Kidwell and Lisch, 2001). They are more ubiquitous in eukaryotes where they are present in virtually all species investigated to date, with few exceptions (Wicker *et al.*, 2007). In prokaryotes, on the other hand, more than 20% of sequenced genomes lack TEs or their remnants (Touchon and Rocha, 2007). TEs also tend to be more abundant in eukaryote genomes, making up to 80% of the genome. For example, they comprise 60% of the maize genome (Messing and Dooner, 2006), 45% of the human genome (Cordaux and Batzer, 2009; Lander *et al.*, 2001), and 15% of the *D. melanogaster* genome (Dowsett and Young, 1982), while in prokaryotes, they form only a maximum of 10% of genomes (Hua-Van *et al.*, 2011).

Thomas (1971) famously coined the term “C-value paradox” to define the then-curious lack of correlation between genome size (measured as DNA content or C-value) and the biological complexity of eukaryotes. Subsequently, it was found that, rather than correlating with gene content, genome size often correlates with quantities of TE and TE-derived DNA. In fact, because the abundance of TEs within a genome can vary widely (Biémont and Vieira, 2006), they, in addition to repetitive DNA, are major determinants of genome size within taxa (Bennetzen, 2005). For example, the genome size of barley is 10 times larger than that of rice (Argumuganathan and Earle, 1991), a related grass with which it shows a great degree of synteny except that its genes are separated by large clusters of retrotransposons.

III. EFFECTS OF TES ON HOST FITNESS AND EVOLUTION

A. TEs as selfish DNA

Selfish genetic elements (SGEs) may be defined as stretches of DNA that act narrowly to advance their own interests at the expense of the whole organism by ensuring that a disproportionate fraction of offspring carry the DNA in question (Burt and Trivers, 2006). The concept of TEs as "selfish" or "parasitic" was codified in seminal papers by Doolittle and Sapienza (1980) and Orgel and Crick (1980), but while the view of TEs as SGEs is now widely accepted (Werren, 2011), it perhaps obscures the continuum of interactions (from extreme parasitism to obligate mutualism) between host and TE (Kidwell and Lisch, 2001) that often profoundly influence host genome evolution (Biémont, 2010; Biémont and Vieira, 2006; Feschotte, 2008; Hua-Van *et al.*, 2011; Hurst and Werren, 2001; Kidwell and Lisch, 2001). Nevertheless, it is clear that the default view of TEs cannot be that they are simply functional parts of the genome. Brookfield (2005), in developing an analogy first made by Kidwell and Lisch (2001), describes the interaction of TEs and their hosts (and indeed between TEs within a host) in terms of the "ecology of the genome." He suggests that questions about TE numbers, diversity, and population dynamics within genomes have ecological parallels with species in communities, and ecology therefore provides insights into the biology of TEs.

While most SGEs compete for representation at a single locus, TEs accumulate by copying themselves to new genomic locations and it is this unique aspect of their drive that lies at the heart of their influence on host fitness and evolution. Because TEs can transpose at a frequency (typically 10^{-5} to 10^{-3} per element per generation) that is often much higher than classical nucleotide-base substitution rates (10^{-9} to 10^{-8}), they are powerful producers of the raw material for evolution (Biémont and Vieira, 2006). The mutations caused by TE insertion and excision are also diverse, encompassing a broad spectrum from small-scale nucleotide changes to large chromosomal rearrangements (Hua-Van *et al.*, 2011; Kidwell and Lisch, 2001) including TE-mediated gene duplication (Jiang *et al.*, 2004; Xiao *et al.*, 2008; Yang *et al.*, 2008). The combination of these two factors means that TEs may play an especially important role in evolution as the main source of spontaneous internal mutations (Kidwell and Lisch, 2001; Li *et al.*, 2007). For example, the high rate of new insertions of *Alu* and *LINE-1* elements (Xing *et al.*, 2009) means that TE insertions are a significant source of mutations in humans (Cordaux and Batzer, 2009). Additionally, 50–80% of mutations in *Drosophila* are the result of TE insertions (Biémont and Vieira, 2006; Finnegan, 1992; Green, 1988).

B. Negative effects on hosts

As with other types of mutation, TE-induced changes will tend to be either harmful or neutral in their fitness effects on the host. TEs harm hosts in a number of ways. Insertions may disrupt coding sequences or *cis*-regulatory regions, while recombination between TE copies can result in deletions and rearrangements. On top of this are the costs to the host of transcription and translation of large numbers of TEs (Charlesworth *et al.*, 1994; Kidwell and Lisch, 2001; O'Donnell and Burns, 2010). Fitness reductions have been quantified for *P* element transposition in *D. melanogaster* (e.g., Currie *et al.*, 1998; Fitzpatrick and Sved, 1986; Mackay, 1986, 1989; Mackay *et al.*, 1992) where even nonlethal inserts tend to reduce host fitness by as much as 12.2% per insert when homozygous (Eanes *et al.*, 1988; Mackay *et al.*, 1992).

In most cases, highly deleterious insertions will be quickly removed by selection, but areas of the genome which experience low recombination might be expected to accumulate insertions that have even moderately harmful effects. Y chromosomes and neo-Y chromosomes, where TE fixation rates tend to be much higher than on X chromosomes or autosomes, present such a case—since recombination is suppressed, selection is expected to be less effective due to hitchhiking and other effects (Charlesworth and Charlesworth, 2000). For example, TEs have accumulated at a very high abundance on the *Drosophila miranda* neo-Y chromosome and might have been involved in causing a loss of gene activity (Steinemann and Steinemann, 1998).

C. TE population dynamics

While there is continuing debate as to which of the various sources of harm are more important (Burt and Trivers, 2006), it is the interplay between selection for increased replication at the TE level, but against deleterious host fitness effects that is responsible for TE population dynamics. Most of the deleterious phenotypic effects of TEs will be removed from a population over time by purifying selection (Kidwell and Lisch, 2001). Nonetheless, a population genetics model has shown that TEs can produce significant deleterious effects in the host and still spread in the population (Hickey, 1982). Other models (e.g., Brookfield and Badge, 1997) highlight the importance of host population demography on TE copy number. In these models, factors such as small host effective population size (N_e) attenuate the power of natural selection in regulating TE copy number. Empirical evidence suggests that these factors play an important part in TE copy number and distribution in natural populations (Charlesworth and Charlesworth, 1995; Lockton and Gaut, 2010; Lockton *et al.*, 2008). Lynch and Conery (2003) suggest that many aspects of complex genomes such as TE abundance were indirect consequences of reduced N_e, producing less effective selection against mildly deleterious insertions.

While transposition rates tend to exceed excision rates, there is strong evidence that TE copy number is regulated. For example, the most abundant TE family still active in *D. melanogaster* is the retrotransposon *roo*, and there are only 60 full-length copies per haploid genome in the euchromatin (Kaminker *et al.*, 2002). Charlesworth and Charlesworth (2010) list five kinds of processes which may be involved in regulating TE abundance: (1) self- and/or host regulation of transposition rates, (2) selection against mutations, (3) ectopic exchange, (4) direct negative fitness effects of transposition on host fitness, and (5) indirect effects of copy number on fitness. They conclude that while each of these processes is plausible, and they are not mutually exclusive, the ectopic exchange model seems to be most consistent with current evidence (Charlesworth and Charlesworth, 2010; Charlesworth *et al.*, 1997). Petrov *et al.* (2011) come to similar conclusions while examining the population frequencies of 755 TEs in six *D. melanogaster* populations.

D. Negating host fitness costs

1. Cost minimization at the level of TEs

The fate of a TE in its host population thus depends not only on transposition rate but also on host fitness effects, and TEs themselves should evolve to reduce host harm (Burt and Trivers, 2006). Germ line specificity of transpositional activity, as demonstrated in a number of class I (e.g., *I* elements and *gypsy*) and class II (e.g., *P* elements and *hobo*) elements in *Drosophila*, is one such adaptation (Burt and Trivers, 2006) since transposition within the soma does not benefit the TE but does damage the host (Charlesworth and Langley, 1986).

Another damage-limiting strategy adopted by TEs is to insert preferentially into safe sites in the genome, as seen in *Ty1*, *Ty2*, *Ty3*, and *Ty4* retrotransposons in baker's yeast which target intergenic regions upstream of tRNA genes (Kim *et al.*, 1998). Additionally, there are many TEs which integrate into gene-rich regions, but which use mechanisms that prevent the disruption of open-reading frames (ORFs) (Levin and Moran, 2011). One example of this is seen in *D. melanogaster P* elements which tend to avoid disrupting ORFs by inserting within 500 bp upstream of host gene transcription start sites (Bellen *et al.*, 2011). Other safe haven transpositions include insertion into other TEs and preferential insertion at or near telomeric chromosome ends. Examples of the latter include the *HeTA*, *TART*, and *TAHRE* non-LTR retrotransposons which comprise the ends of *D. melanogaster* chromosomes (Biessmann *et al.*, 1992; George *et al.*, 2010; Levis *et al.*, 1993).

It has also been proposed that some TEs may have evolved autoregulation of transposition rate to avoid the deleterious effects of uncontrolled transposition bursts (Burt and Trivers, 2006; Hua-Van *et al.*, 2011). Theory suggests that the circumstances under which such regulation would evolve are probably common, although unlikely to exist in unstructured random-mating hosts (Charlesworth and Langley, 1986, 1989). Nevertheless, there are examples where self-regulation appears to be the case such as the *P* element-encoded repressor which represses transposition and excision (Robertson and Engels, 1989).

2. Cost minimization at the level of the host: TE suppression

Hosts are not defenseless against harmful transposition. Many organisms have evolved complex mechanisms to deal with TEs. Small RNA-based mechanisms act to defend eukaryotic cells against TEs by posttranscriptional disruption of TE mRNA (Aravin *et al.*, 2007; Malone and Hannon, 2009; van Rij and Berezikov, 2009). Another way in which some host taxa suppress their TEs is through epigenetic control, including methylation. In fact, it is widely thought that epigenetics, whose processes are commonly used by metazoans in cell lineage-specific gene regulation, first evolved to defend against foreign DNA including TEs (Hua-Van *et al.*, 2011). This is one example of how the prolonged interaction of host and TE has ultimately benefited the host—it is far from the only one.

E. Beneficial effects of TEs

As with any other source of mutation, TEs can occasionally produce beneficial genetic alterations to host genomic DNA. A beneficial insertion would be expected to go to fixation within a population, and TE fixation has been observed, particularly in *D. melanogaster* (González and Petrov, 2009). The *S* element(s) associated with the Hsp70 (heat-shock protein) genes in *D. melanogaster* is one possible example of a beneficial TE (Maside *et al.*, 2002). While the functional significance of this insertion has not been elucidated, there is strong evidence of a selective sweep around it. Furthermore, the insertion apparently occurs in a freely recombining region of the genome, which substantially lowers the probability of fixation via drift.

F. Co-option/domestication

In contrast to the benefits derived from genetic alteration of host genomic sequences *per se*, TE sequences themselves may be co-opted for host function, a process which has been called "domestication" or "exaptation" when TE-coding sequence function has been appropriated for host use. There are several examples of this in the literature, one of the most cited being the full domestication of the *Drosophila* telomeric retrotransposons *HeTA* and *TART* which function as telomerase to heal chromosome ends. Even noncoding TE sequences may be useful—one striking example of the fixation of a beneficial insertion which is of particular importance to this review is found in the evolution of DDT resistance in *D. melanogaster*. Here, an *Accord* retrotransposon-derived sequence inserted upstream of a cytochrome P450 gene has been shown to upregulate the detoxification enzyme and increase pesticide resistance (Chung *et al.*, 2007; Daborn *et al.*, 2002). The remainder of the review focuses on insecticide resistance and how TEs influence this.

IV. INSECTICIDE RESISTANCE

A. The rate of evolution

Given the evolutionary potential of TEs, perhaps it is not surprising that they play an important role in such key fitness traits as pesticide resistance. Over the past 100 years, there has been an increased use of toxic chemicals to control pest organisms, particularly from the 1950s onward (Wilson, 2001). This strong, pervasive source of selection has demonstrated the tremendous capacity of populations to evolve resistance to toxins. Since the first insecticide resistance case was reported almost a century ago (Melander, 1914), there have been thousands of cases of resistance in hundreds of species (Georghiou and Lagunes-Tejeda, 1991; Whalon *et al.*, 2008). Some of the most dramatic examples of microevolution in action have come from selection for chemical resistance (Hartl and Clark, 1997), with resistance evolving in as few as 5–50 generations (May, 1985) and toward rapid global fixation in many insect pest populations (Catania *et al.*, 2004; Schlenke and Begun, 2004; Whalon *et al.*, 2008).

B. What is resistance?

From a functional point of view, insecticide resistance may be defined as the ability of an organism to survive a dose of insecticide that is lethal to a susceptible one (Georghiou and Saito, 1983), and dynamically, it has also been described as the microevolutionary process whereby genetic adaptation through pesticide selection results in populations of susceptible insects being replaced by resistant ones over a period of time (Wilson, 2001). The biochemical mechanisms and molecular genetics underlying resistance have been well studied and have been the subject of several books (Clark and Yamaguchi, 2002; Denholm *et al.*, 1999; Ishaaya, 2001) and reviews (Feyereisen, 1995; ffrench-Constant *et al.*, 2004; Oakeshott *et al.*, 2003).

The proximate biochemical mechanisms of resistance can be divided into four main categories (Wilson, 2001). The first of these is behavioral resistance (i.e., avoidance of the insecticide), which may involve genetic changes, but is probably of minor importance even though it has been documented for a few species (Sparks *et al.*, 1989). Reduction in the penetrative ability of the toxin is a second mechanism, but again this does not seem to be of major importance (Wilson, 2001). Target-site inactivation (changes in the insecticides site of action) is a very important biochemical resistance mechanism (Hollingworth and Dong, 2008; Wilson, 2001). Every potent insecticide has one or more specific binding sites on critical macromolecules, and changes in the ability of the toxin to bind must affect its impact on the insect (Hollingworth and Dong, 2008). Lastly, biotransformation, the metabolic breakdown of a toxin, is a common defense against natural xenobiotics (Li *et al.*, 2007). It is therefore not surprising that, with the widespread use of synthetic organic agricultural chemicals, the enzymatic systems which originally evolved to detoxify phytotoxins should been enlisted to defend against insecticides (Wilson, 2001). Three types of enzymes—esterases (through ester hydrolysis), cytochrome P450 monoxygenases (through oxidation), and glutathione transferases (through ester hydrolysis)—are commonly used to transform insecticides into less toxic products (Hollingworth and Dong, 2008).

When an insecticide is first introduced, the target population largely consists of susceptible phenotypes (Macnair, 1991; Mallet, 1989; McKenzie and Batterham, 1994; Roush and McKenzie, 1987). Within the population, there will be a distribution of susceptibility based on factors such as size, age, and physiological condition (McKenzie and Batterham, 1994), which are generally polygenically inherited. Insecticide selection on this distribution will act via the phenotype and resistance will be polygenically inherited, combining preexisting factors of primarily minor effect (such as size and developmental rate) (ffrench-Constant *et al.*, 2004). This type of selection is seen in most laboratory studies (ffrench-Constant *et al.*, 2004; McKenzie and Batterham, 1994), which explains why early studies of DDT resistance (e.g., Crow, 1957) determined that resistance evolution was a polygenic response.

This contrasts strongly with what has been found in natural populations, where resistance to particular insecticides often involves one or two major genes (ffrench-Constant *et al.*, 2004; Field *et al.*, 1988; Mallet, 1989; McKenzie and Batterham, 1994; Raymond *et al.*, 1989; Roush and McKenzie, 1987). This may represent detection bias, but another explanation could be that insecticides in the field tend to occur at concentrations which favor variation outside of the normal phenotypic distribution (i.e., rare resistant mutations of major effect). Natural populations are much larger than laboratory populations and so more likely to contain individuals with these rare mutations. A second reason for the preponderance of monogenic resistance in the wild may be evolutionary constraint resulting from opposing natural selection on multiple targets. The nature of the ultimate genetic changes which lead to monogenic resistance also varies with respect to the proximate biochemical mechanism involved (Wilson, 2001).

C. Mechanisms

Target-site inactivation is usually effected by subtle changes in the target protein—it is therefore easy to understand the importance of point mutations for this resistance mechanism (Wilson, 2001). An altered protein must retain at least some degree of normal function while decreasing its xenobiotic sensitivity, which explains the highly conserved nature of such changes (ffrench-Constant, 1999; ffrench-Constant *et al.*, 1998; Li *et al.*, 2007; Wilson, 2001). A striking illustration of this is the parallel evolution of cyclodiene resistance in a wide range of pest species and in *Drosophila*, which is a result of the same single amino acid substitution in the chloride ion channel pore of the gamma-aminobutyric acid receptor protein (ffrench-Constant *et al.*, 1998; Thompson *et al.*, 1993).

Metabolic resistance, on the other hand, tends to involve the overexpression of existing metabolic enzymes either through gene amplification (i.e., gene duplication, which results in more gene product) or alterations in their regulatory systems, which increase transcription and/or stabilize mRNA (Hollingworth and Dong, 2008; Li *et al.*, 2007; Wilson, 2001). Examples of resistance through gene copy increase are seen for esterase genes in mosquitoes and aphids, GSTs in the housefly and the aphid *Nilaparvata lugens* and cytochrome P450s in three dipterans including *D. melanogaster* and *D. simulans* and the potato aphid *Myzus persicae* (reviewed in Bass and Field, 2011; Devonshire and Field, 1991). A particularly striking example is provided by resistant *Culex pipiens quinquefasciatus* mosquitoes, where the esterase gene *B1* is amplified in a tandem array as much as 250-fold, conferring high organophosphate (OP) resistance (Karunaratne *et al.*, 1993; Mouchès *et al.*, 1986, 1990).

Gene upregulation is the most common process involved in P450-mediated insecticide resistance, but upregulation has also been documented for the other two major classes of detoxification enzymes already mentioned (Li *et al.*, 2007). This is usually achieved through changes (point mutations or indels) in either *cis*- or *trans*-regulatory loci. An example of the former is provided by the P450 *Cyp6g1* gene in *D. melanogaster* where the insertion of a defective copy of the *Gypsy*-like LTR retrotransposon *Accord* in the 5′ promoter region results in upregulation of the enzyme and cross-resistance to DDT, imidacloprid, nitenpyram, and lufenuron (Catania *et al.*, 2004; Daborn *et al.*, 2002; Schlenke and Begun, 2004). As an example of the latter, overexpression of a GST allele in the resistant *Aedes egypti* GG strain is due largely to a loss-of-function mutation in an unidentified *trans*-acting repressor that represses mRNA transcription and/or decreases mRNA stability in the susceptible strains (Grant and Hammock, 1992).

D. Costs of resistance?

A central question in the evolution of resistance is the fitness of the organism carrying a mutant allele of a resistance gene. Theory holds that, in the absence of insecticide, the majority of insecticide-resistant organisms should show some differential survival in comparison with "wild-type" organisms. That is, resistance should be costly (e.g., Crow, 1957). However, empirical evidence on the pleiotropic fitness effects of insecticide resistance appears to be equivocal. There are a few empirical studies that confirm that investment in resistance entails a fitness cost (Alyokhin and Ferro, 1999; Berticat *et al.*, 2002; Boivin *et al.*, 2001; Carrière *et al.*, 1994, 1995, 2001; Chevillon *et al.*, 1997; Foster *et al.*, 2003; Minkoff and Wilson, 1992; Rivero *et al.*, 2011; Smith *et al.*, 2011; Yamamoto *et al.*, 1995). On the other hand, some authors have failed to reveal any detrimental effects of insecticide resistance (Baker *et al.*, 1998, 2008; Castañeda *et al.*, 2011; Follett *et al.*, 1993; Tang *et al.*, 1997, 1999), and some have demonstrated pleiotropic fitness benefits (Arnaud and Haubruge, 2002; Bielza *et al.*, 2008; Bloch and Wool, 1994; Haubruge and Arnaud, 2001; Mason, 1998; McCart *et al.*, 2005; Omer *et al.*, 1992; White and Bell, 1995).

In other studies, some measures of fitness have been negatively affected, others positively (Brewer and Trumble, 1991), and this may even involve sexual antagonism, where resistance alleles have opposing fitness effects depending on which sex they reside. This has recently been documented for DDT resistance in *D. melanogaster*, where resistance confers a strong fecundity advantage to females, but a competitive mating disadvantage to males (McCart *et al.*, 2005; Smith *et al.*, 2011). In addition, how resistance alleles impact nonresistance-related fitness can depend on the strain being investigated (Chevillon *et al.*, 1997; Hollingsworth *et al.*, 1997; Oppert *et al.*, 2000; Smith *et al.*, 2011). This reflects epistasis, where the pleiotropic fitness effect is mediated by the genotype (or genetic background) of the insect in question.

V. TES CONFERRING INSECTICIDE RESISTANCE

A. Initial findings

Wilson (1993) was the first to speculate that TEs were implicated in insecticide resistance, although the evidence was indirect—he was able to generate Methoprene-resistant alleles in *D. melanogaster* using *P* element mutagenesis (Wilson and Turner, 1992). Around the same time, Waters *et al.* (1992) found an association between *Drosophila* strains resistant to DDT and Malathion and a *17.6* TE insertion in the 3′ region of a cytochrome P450 enzyme gene. In this case, it was found that the resistant strains lacked the insertion, suggesting that resistance was a result of an excision of the TE. However, Delpuech *et al.* (1993) subsequently reported that the presence or absence of the *17.6* LTR was uncorrelated with resistance in 31 strains of *D. melanogaster* and *D. simulans*.

Wilson (2001) was less convinced about the possibility that TEs play a significant role in insecticide resistance in nature (notwithstanding the *P* element-induced resistance, he demonstrated in the laboratory a decade earlier) conceding that, at most, "TE mutagenesis may be important only for a few genes where resistance can result from severe underexpression or nonfunctional gene product." However, since his review, evidence has been steadily accumulating that TEs do, in fact, play an important part in the evolution of insecticide resistance.

The observation that TEs are frequently found within or in close proximity to resistance genes provides indirect evidence that TEs are involved in resistance-related adaptive genomic changes (Li *et al.*, 2007). This inference was bolstered by the findings of Chen and Li (2007) who reported that TE insertions were enriched around and within xenobiotic-metabolizing P450 genes of both *Helicoverpa zea* moths and *D. melanogaster* flies. They also found that TE insertions were absent from essential housekeeping P450 genes in *D. melanogaster*, which might be expected since mutation of essential genes is more likely to be lethal and not simply reduce fitness. Taken together, these results indicate that TEs are also selectively retained within or in close proximity to xenobiotic-metabolizing P450 genes. Similarly, while a *Bari-1* element insertion occurs downstream of the P450 gene *Cyp12a4* in an Australian lufenuron-resistant *D. melanogaster* strain, its presence in lufenuron-susceptible strains suggests that while the insertion may be important, it is not the main cause of resistance (Bogwitz *et al.*, 2005).

Recent studies provide more conclusive, direct evidence for a causative link between resistance and TEs. For example, insertion of a 2.3-kb LTR retrotransposon *Hel-1* in the putative Bt-toxin receptor gene cadherin leads to 3′-truncated nonfunctional cadherin protein and Bt resistance in a laboratory-selected *Heliothis virescens* strain (Gahan *et al.*, 2001). Furthermore, parallel insertions of *Accord*-LTR or *Doc* non-LTR retrotransposon into the 5′-regulatory region of *Cyp6g1* in *D. melanogaster* or *D. simulans* are associated with *Cyp6g1* upregulation and DDT resistance (Daborn *et al.*, 2002; Schlenke and Begun, 2004) (Section IV.D). Additionally, in *D. melanogaster*, insertion of a *Doc1420* retrotransposon into the second exon of the predicted gene CG10618 (*CHKov1*, a putative choline kinase gene) generates two sets of altered transcripts and a novel polypeptide (Aminetzach *et al.*, 2005). Whether through loss of original *CHKov1* function or through function of the new protein, the *Doc1420* insertion confers moderate OP resistance (Aminetzach *et al.*, 2005).

B. Why are TEs so important?

Insecticide resistance results from very strong, persistent directional selection. TE-mediated changes in regulation can lead to massive and rapid changes in expression, responses that are potentially highly adaptive when an organism is faced with a major, pervasive, and novel mortality agent in the environment, like an insecticide. A useful contrast which illustrates this point is the essential absence of TEs involved in natural xenobiotic resistance—if we consider that mutational changes in plant allelochemicals are unlikely to bring about massive changes in mode of action or in toxicity, then mutational change associated with allelochemical resistance may be acquired more slowly as a result of the accumulation of small changes in structural genes (Li *et al.*, 2007).

Application of insecticide tends to favor insecticide resistance, involving single genes of major effect rather than polygenic resistance (ffrench-Constant *et al.*, 2004), and it has been found that most resistant field strains show monogenic resistance (Roush and McKenzie, 1987). Where resistance genes are already involved in essential functions, as is often the case for metabolic enzymes, it is advantageous to maintain the quality of mRNA to allow wild-type function to be retained and instead regulate gene expression. TE insertion within regulatory regions of genes which confer resistance often results in upregulation, that is, increase in the quantity of mRNA. This may be because many TEs have built-in enhancer sequences related to their transposition (Zhang and Saier, 2009) that have been co-opted by the host, but another possibility is that such spacing may move genes further from existing regulatory sequences (Schlenke and Begun, 2004).

C. Mechanisms of resistance via TEs

ffrench-Constant *et al.* (2006) list four possible mechanisms whereby TE insertions might confer insecticide resistance. First, a TE insertion in the 5′-end of a gene may introduce a novel enhancer sequence. The *Accord*-LTR upstream of the cytochrome P450 gene *Cyp6g1* in *D. melanogaster* is one such case (Chung *et al.*, 2007), and the *Cyp6g1* homolog in *D. simulans*—where the insertion is a complete *Doc* element—may also be one (Schlenke and Begun, 2004). Another example of this mechanism is found in the mosquito *Culex quinquefasciatus* where the insertion of a miniature-inverted terminal repeat (MITE)-like element upstream of another cytochrome P450 gene is associated with increased pyrethroid resistance (Itokawa *et al.*, 2010). The second mechanism involves increased mRNA stability via TE insertion in the 3′-end of a gene which increases the final pool of translatable RNA. Third, TEs might excise a gene and move it to a different genomic location away from local repressor elements normally responsible for shutting off expression or to a position proximal to an enhancer element. This position effect was demonstrated in principle by Berrada and Fournier (1997) who used *P* element-mediated transposition to initiate transcriptional overexpression of an artificially constructed acetylcholinesterase minigene in *D. melanogaster*. Finally, TE insertion might alter the pattern of resulting transcripts and potentially lead to a truncated gene product of novel function as appears to be the cases described by Gahan *et al.* (2001) and Aminetzach *et al.* (2005). One further mechanism not mentioned by ffrench-Constant *et al.* (2006) involves gene amplification. The transpositional mechanism of TEs may result in gene duplication through ectopic recombination (e.g., Yang *et al.*, 2008) and consequent increase in gene product, or evolution of new gene function in the duplicated gene.

D. TE-mediated DDT resistance in *D. melanogaster* and *D. simulans*

Fifty years ago, studies in *D. melanogaster* indicated that many genes contributed to DDT resistance (Crow, 1957; Dapkus and Merrell, 1977; Hallstrom, 1985; Kikkawa, 1961) including loci on all three major chromosomes. However, work on the Hikone-R strain indicated that resistance in this strain was largely conferred by a single dominant locus on chromosome II, and this was later found to be *Cyp6g1* (Daborn *et al.*, 2001). Resistant alleles were found to have a defective copy of an *Accord*-LTR retrotransposon inserted about 300 bp upstream of the transcription start site. Subsequently, the molecular mechanism of upregulation was identified (Chung *et al.*, 2007) with *cis*-regulatory sequences in the *Accord*-LTR being responsible for increased *Cyp6g1* transcription.

Schlenke and Begun (2004), while investigating reduced heterozygosity around the *Cyp6g1* locus in *D. melanogaster* and *D. simulans*, found that another TE insertion, this time a full-length copy of the non-LTR retrotransposon *Doc*, occurred 200 bp upstream of the gene in Californian populations of the latter species. Once again, the insertion correlated with increased *Cyp6g1* expression compared with that found in African populations lacking the insertion. In contrast to the *Accord* insertion in *D. melanogaster* which is highly degenerate (comprising only the LTR), the *Doc* insertion in *D. simulans* is of an autonomous element, suggesting that it is a much more recent event. Selective sweeps at *Cyp6g1*, associated with strong recent selection, were demonstrated in both species (Catania *et al.*, 2004; Schlenke and Begun, 2004).

Catania *et al.* (2004) conducted a survey of the *Accord*-LTR insertion at *Cyp6g1* in 673 lines from 34 populations from around the world. They found near fixation of the *Accord*-LTR-inserted alleles in non-African and North and Western African *D. melanogaster* populations (85–100% of chromosomes sampled), with significantly lower frequencies in East African populations (32–55%). Variation in the *Accord*-LTR-inserted allele was also found—diagnostic PCR revealed some variability in product size, and subsequent cloning and sequencing of the variants revealed an insertion of a partial *P* element nested within the *Accord*-LTR in a New Delhi line (Catania *et al.*, 2004). Furthermore, Emerson *et al.* (2008), using genome-wide tilling arrays, found copy number polymorphism in *D. melanogaster* at *Cyp6g1*, with 13 of 15 lines tested showing a duplication encompassing both *Cyp6g1* and *Cyp6g2*.

Most recently, Schmidt *et al.* (2010) characterized copy number variation and further allelic variation at the *Cyp6g1* locus. Characterization of *Cyp6g1* copy number variation and TE insertion complexity in the *D. melanogaster* RK146 strain revealed two full-length copies of *Cyp6g1*, named *Cyp6g1-a* and *Cyp6g1-b*. A repeat unit was found between the two full-length copies that contained a fusion of partial copies of both *Cyp6g1* and *Cyp6g2*, the gene found downstream of *Cyp6g1-b*. Both copies of *Cyp6g1* were found to contain the LTR of the *Accord* element. Unexpectedly, a *HMS-Beagle* TE was found inserted into the *Accord*-LTR upstream of *Cyp6g1-a*. Testing of other *D. melanogaster* lines revealed that the partial *P* element found by Catania *et al.* (2004) was located upstream of *Cyp6g1-b*.

Schmidt *et al.* (2010) also demonstrated an allelic progression of five different alleles, including the ancestral allele lacking any TE insertions and four alleles that involve duplication of *Cyp6g1* (Fig. 2.2). They reason that these alleles represent multiple adaptive steps at *Cyp6g1*, with increased DDT resistance being demonstrated along the allelic progression. Their survey of *D. melanogaster* global populations showed that most flies in Europe, Asia, and the United States carry the *Cyp6g1* duplication and the double *Accord*-LTR (no *P* element) insertion or the *Cyp6g1* duplication and combined *Accord*-LTR insertion/*Accord*-LTR (with nested *HMS-Beagle*) insertion.

The question of the age of the original *Accord*-inserted allele remains open. In their survey, Schmidt *et al.* (2010) did not find any *Cyp6g1* alleles with the insertion that did not also represent a duplication, nor did they find any gene duplicates which did not also contain the *Accord* insertion. This strongly suggests that either the original insertion and duplication events occurred simultaneously, or, more likely, the original *Accord* insertion preceded the duplication. If the latter, then we are yet to find the original *Accord*-inserted allele.

Catania *et al.* (2004) suggested that the lower than expected reduction in variability in microsatellites around the *Cyp6g1* locus could be explained most parsimoniously if the *Accord*-LTR insertion occurred at low frequency in African populations before the species' global expansion. This would imply that the insertion was already part of the genetic variation at this locus well before it permitted adaptation to insecticide and may be an example of how TEs provide latent genetic variation facilitating adaptive responses to selection.

The absence of strong DDT selection since its ban in most countries globally over the past 30 years has not brought about the loss of DDT-R alleles in *D. melanogaster*, as might be expected from theory—overexpression of P450

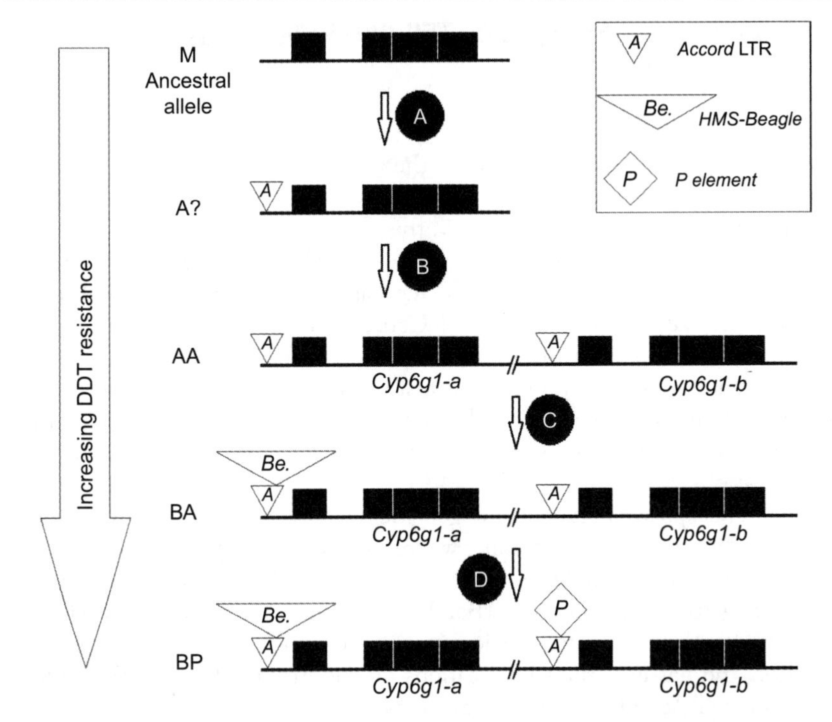

Figure 2.2. Allelic progression at the *Cyp6g1* locus in *D. melanogaster* as described by Schmidt *et al.* (2010). The ancestral allele (*Cyp6g1-[M]*) is found in DDT-susceptible strains such as Canton-S and is the only allele found in laboratory lines established in the 1930s. The most plausible sequence of changes to the ancestral allele are as follows: (A) Insertion of an *Accord* retrotransposon 5′ of *Cyp6g1*, followed by excision (leaving only the LTR footprint), produces hypothetical allele *Cyp6g1-[A?]*. (B) Duplication of the *Accord*-LTR-inserted allele produces a tandem repeat of two full-length *Cyp6g1* copies (*Cyp6g1-a* and *Cyp6g1-b*) separated by partial *Cyp6g1*/*Cyp6g2* repeat units and resulting in the DDT-resistant allele *Cyp6g1-[AA]*. (C) Insertion of an *HMS-Beagle* retrotransposon into the *Accord*-LTR found in *Cyp6g1-a* of *Cyp6g1-[AA]* produces the DDT-resistant allele *Cyp6g1-[BA]*. (D) Insertion of a *P-element* DNA transposon into the *Accord*-LTR found in *Cyp6g1-b* of *Cyp6g1-[BA]* produces the DDT-resistant allele *Cyp6g1-[BP]*.

genes must have a cost, and if the resistance-to-DDT benefit to balance this cost is not there, then selection should remove the resistance allele. The near fixation of *Accord*-inserted *Cyp6g1* alleles in worldwide populations (Catania *et al.*, 2004) and the seemingly adaptive, ongoing elaboration of these alleles through subsequent TE insertion increasing DDT resistance (Schmidt *et al.*, 2010) thus poses a puzzle. The most obvious explanation is that these alleles confer cross-resistance to other pesticides and therefore remain under strong xenobiotic selection. There are other explanations, however.

VI. SEX-SPECIFIC EFFECTS OF TES INDEPENDENT OF DDT RESISTANCE

The strong fitness benefit conferred by an *Accord*-inserted *Cyp6g1* allele to Canton-S strain females and pupae in the absence of pesticide (McCart *et al.*, 2005) offers another explanation for the persistence of these resistance alleles at high frequencies in the wild. However, this nonresistance fitness benefit should not be viewed in isolation—Smith *et al.* (2011) very recently confirmed a competitive mating disadvantage to males carrying the same allele and in the same genetic background as used by McCart *et al.* (2005). This male fitness cost almost perfectly balances the benefit conferred to females. The presence of a DDT-R male fitness disadvantage was first suggested by Drnevich *et al.* (2004), who in breeding *D. melanogaster* males of high and low MCRS (male competitive reproductive success) discovered that high MCRS was associated with low expression of *Cyp6g1*.

This apparent sexually antagonistic selection at *Cyp6g1* is an excellent example of intralocus sexual conflict where the precise locus has been identified. Intralocus sexual conflict occurs when an allele has positive fitness effects in one sex and negative in the other sex (Bonduriansky and Chenoweth, 2009; Hosken *et al.*, 2009). Sexually antagonistic selection generally helps maintain genetic variation and has important implications for the history of the DDT-R allele. If the original *Accord*-inserted allele antedates insecticide use, as Catania *et al.* (2004) posit, sexually antagonistic selection may have accounted for its low frequency in populations prior to DDT's introduction—no *Accord*-inserted *Cyp6g1* alleles have been found from lines established in the 1930s (Schmidt *et al.*, 2010).

It should be noted that the male fitness disadvantage of the *Accord*-inserted allele is not consistent across different genetic backgrounds. When the allele was introgressed into another *D. melanogaster* strain, males were still competitively disadvantaged, but not significantly so (Smith *et al.*, 2011). As yet, the fitness effects of the *Accord* insertion in females of the latter genetic background have not been examined. This latter strain had been established by Trudy MacKay in North Carolina, USA, from wild-caught flies in 2004 and contrasts with the Canton-S background which represents one of the earliest established laboratory populations (1930s). The lack of a significant male disadvantage in the more recently established strains may thus represent the evolution of male-specific modifiers in populations dominated by resistant *Cyp6g1* alleles (although this could also represent a statistical power issue). In this context, it is possible that DDT (and other insecticide) selection may have prompted not only rapid spread of the resistance allele but also male-specific counter-adaptation to reduce pleiotropic fitness costs of resistance. Interestingly, while homozygous resistant Canton-S males were significantly smaller than their susceptible counterparts, the converse was true for males of the wild-type genetic background (Smith *et al.*, 2011), hinting that amelioration of the male fitness cost may involve loci influencing male body size.

Cohan *et al.* (1994) list two general ways in which pleiotropic fitness costs of an adaptive mutation may be ameliorated. The epistatic male DDT-R cost effect seen by Smith *et al.* (2011) may be an example of the "compensatory" mode, in which natural selection favors modifiers (at other loci) that compensate for the deleterious effects of the mutant allele (Fisher, 1928). The other mode, known as "replacement," describes the case where there are multiple mutations which confer the same adaptation, but which vary in their pleiotropic fitness costs such that the original mutation is replaced by one which confers the same adaptive benefit at lower cost (Haldane, 1932). Interestingly, this mode has been invoked to explain temporal allele replacement observed in the insecticide resistance gene *Ester* in the mosquito *C. pipiens* in southern France (Labbé *et al.*, 2009). A similar scenario may exist in the *Cyp6g1* allelic progression described by Schmidt *et al.* (2010).

VII. ONGOING AND FUTURE RESEARCH

In spite of observed female fitness benefits in the absence of DDT (McCart *et al.*, 2005), recent simple models (our unpublished data) demonstrate that an *Accord*-inserted *Cyp6g1* (i.e., DDT-R) allele could have been kept at low frequency in populations before DDT was introduced, through male-associated fitness costs (Smith *et al.*, 2011). The net effect of male and female pleiotropic fitness components, with parameter values based on empirical work by Daborn *et al.* (2001), McCart *et al.* (2005), and Smith *et al.* (2011), was a very slow return to a stable polymorphism at the *Cyp6g1* locus (Fig. 2.3). These results are therefore consistent with continued high levels of DDT-R several years after the removal of DDT selection, a phenomenon which has been previously explained by invoking cross-resistance, a lack of fitness cost and low migratory rates (Catania *et al.*, 2004; McCart *et al.*, 2005).

Further work is required to determine the range of fitness effects of DDT-R in different genetic backgrounds and to further explore the population genetics of these alleles. We also need to investigate the underlying reasons for the competitive disadvantage observed in resistant males of Canton-S background—is it simply size mediated or are there other mechanisms (behavioral and/or physiological) which affect male fitness? Related to this, the demonstration of epistasis with regard to male fitness effects begs a more thorough investigation of the distribution of pleiotropic fitness effects in both sexes. How common is the DDT-R-associated male competitive disadvantage? How common is the female fitness benefit? In this respect, it would be useful to run

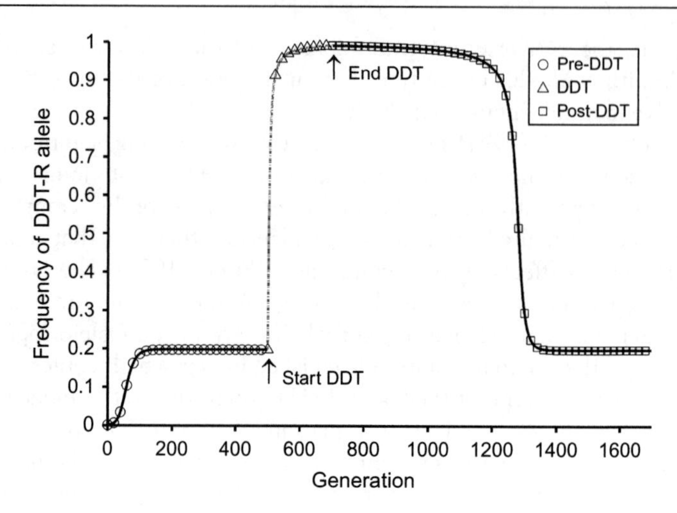

Figure 2.3. Results of a population genetic model showing the changes in DDT-R frequency prior to, during, and after DDT selection. Simulations incorporate the sexually antagonistic fitness effects of the allele in the absence of DDT. The initial allele frequency was set very low (initial DDT-R frequency is 0.001) to demonstrate that the balance of male and female fitness effects can move a rare allele to a stable polymorphism. Note that the fixation of the allele is very rapid when DDT use is introduced and the return to the polymorphic equilibrium after relaxing the strong directional selection for resistance (DDT) is initially slow. Symbols plotted every 20 generations.

population cage studies and observe temporal change in replicate laboratory populations with different initial *Cyp6g1* allele frequencies. This would also complement population genetic models. At the molecular level, there remains an opportunity to further investigate the genetic basis of DDT-R epistasis—what are the modifiers altering the male fitness costs? Although *D. melanogaster* is not a pest species, understanding the relative reproductive success of susceptible and resistant flies with differing genetic backgrounds could provide valuable baseline data to inform insecticide resistance management programs for pest species. Work to date suggests that using a single genetic background to test for effects may not be representative.

Given the great variation recently discovered at the*Cyp6g1* locus in *D. melanogaster* (Schmidt *et al.*, 2010), the time is ripe to investigate the fitness effects of these newly discovered alleles. In the case of the more derived alleles, females have much higher DDT resistance than males (Schmidt *et al.*, 2010), which points toward possible mitigation of the intralocus sexual conflict described in Smith *et al.* (2011). One hypothesis, easily testable through a fitness components approach and/or selection-based stability analysis *sensu* Raymond *et al.* (2011) (population cage experiments described above), is that the allelic succession may be partially driven by the "replacement" mode of amelioration. However, intralocus conflict resolution like this presumably requires a change to genetic architecture, which makes resolution more difficult than it may appear (e.g., Harano *et al.*, 2010).

It is unknown whether gene amplification or TE-induced *cis*-acting mutation has the greater effect on DDT resistance and associated pleiotropic fitness effects in *D. melanogaster*—dissecting the respective contributions to resistance/fitness would require single-copy TE-inserted *Cyp6g1* alleles, and these are yet to be found (Schmidt *et al.*, 2010). The universal presence of TE insertions in both copies of all DDT-R alleles thus found suggests that the insertion occurred prior to, or concurrently with, the duplication event.

This parallels pyrethroid resistance in the mosquito *C. quinquefasciatus*. Here, resistance is associated with overexpression of another cytochrome P450 gene, *Cyp9m10* (Hardstone *et al.*, 2010; Itokawa *et al.*, 2010, 2011). As with *D. melanogaster* DDT-R, the constitutive upregulation occurs in haplotypes that have an upstream insertion of a TE (in this case a truncated copy of the MITE TE, *CuRE1*; Itokawa *et al.*, 2010). Moreover, one of the resistant haplotypes also consists of a tandem repeat of the TE-inserted sequence (Itokawa *et al.*, 2011). Unlike the *D. melanogaster* DDT-R system, the relative contributions of the TE insertion and gene amplification to resistance (and for that matter pleiotropic fitness) can easily be parsed out, since there are haplotypes that possess the former but not the latter. Itokawa *et al.* (2011) suggest that, based on the nonlinear resistance efficacy to *Cyp9m10* expression, the resistance phenotype is disproportionately stronger as a result of the *cis*-acting mutation (the TE insertion) occurring before the duplication event, than if the duplication had preceded the insertion. Such eerily similar stories for two different enzymes, conferring resistance to two different insecticides in two distantly related species, underline the usefulness of intensive study of model insect systems. They also hint at a general pattern—tandem repeats, which are difficult to detect, could be commonly associated with TE insertion-induced insecticide resistance. This remains to be established.

Compared with the extensive work done on the *Accord*-inserted *Cyp6g1* in *D. melanogaster*, little is known about *Doc*-inserted *Cyp6g1* in *D. simulans*. It remains to be seen whether this mutation has a significant and consistent effect on resistance across different strains/genetic backgrounds. Furthermore, no work has been done to examine potential pleiotropic fitness effects of this insertion, much less the presence of epistatic interactions or the possibility of intralocus sexual conflict, as has been demonstrated for *D. melanogaster*. A good first step may be to perform a worldwide survey akin to that of the *Accord*-LTR insertion by Catania *et al.* (2004). This would provide some indication of the geographic range of the *Doc*-inserted allele. Given the evidence for ongoing and rapid adaptation at *Cyp6g1* in *D. melanogaster*, it may well be worth having a closer look at the variation which exists at this locus in *D. simulans*. Just how similar the responses of the two species are to similar selection also remains to be seen.

Another avenue of research involves gene by environment interaction as it relates to fitness costs of resistance in these model systems. The laboratory-based fitness component approach cannot fully encompass the full diversity of environments in which wild populations face selection, and this may be a reason why costs are not always detected—environmental factors such as natural enemies, resource limitation, overwintering, and different host plant have all been shown to increase resistance costs in various taxa (Carrière *et al.*, 2001; Janmaat and Myers, 2005; Raymond *et al.*, 2005, 2007, 2011). Moving population cage experiments outdoors could increase the reality of the stability-selection approach, giving a better reflection of how well resistance genotypes perform under natural conditions.

Just as the genetic background provided by the rest of the genome represents a genetic "environment" in which resistance alleles act, so does the presence of extragenomic DNA, including cytoplasmic endosymbionts. *Wolbachia* is a maternally transmitted intracellular bacterium found in a wide range of arthropods and nematodes (Stouthamer *et al.*, 1999; Werren, 1997). Its relationship with its host ranges from parasitic to symbiotic. At the parasitic end of the spectrum, it can have profound effects on host reproduction, displaying a range of phenotypes from male killing to feminization to cytoplasmic incompatibility (Stouthamer *et al.*, 1999; Werren, 1997). These strategies increase its transmission within a population, often at the expense of its host's fitness—the hallmark of an SGE. *Wolbachia* is found not only in *Drosophila* (where it has undergone a very recent expansion to near fixation in many populations), but also in many other insects including pest species—one recent estimate is that more than 66% of arthropod species harbor *Wolbachia* infections (Hilgenboecker *et al.*, 2008). Given its ubiquity and potentially profound effect on host fitness, *Wolbachia* cannot be ignored when examining pleiotropic effects of resistance. For example, *Wolbachia* has been implicated in directly modifying the cost of insecticide resistance in mosquitoes (Duron *et al.*, 2006). Where insecticide resistance is conferred by a TE, we may find that intergenomic interactions (akin to epistasis) between the TE and intracellular endosymbionts are critical to the population genetics of insecticide resistance alleles.

Although *D. melanogaster* is not a pest species, understanding the relative fitness of susceptible and resistant flies with differing genetic backgrounds and under different environments could provide valuable insights to inform insecticide resistance management programs for pest species. To this end, we urge the use of multiple avenues of investigation that include the laboratory-based, sex-specific fitness component approach, stability-selection experiments, and mathematical modeling to increase our understanding of insecticide resistance dynamics in natural populations.

Acknowledgments

N. W. is funded by a Royal Society Wolfson Research Merit Award and D. J. H. is funded by NERC. We thank Judith Mank for comments on a previous version of this chapter.

References

Alyokhin, A. V., and Ferro, D. N. (1999). Relative fitness of Colorado potato beetle (Coleoptera: Chrysomelidae) resistant and susceptible to the *Bacillus thuringiensis* Cry3A toxin. *J. Econ. Entomol.* **92,** 510–515.

Aminetzach, Y. T., Macpherson, J. M., and Petrov, D. A. (2005). Pesticide resistance via transposition-mediated adaptive gene truncation in *Drosophila. Science* **309,** 764–767.

Anxolabéhère, D., Kidwell, M., and Périquet, G. (1988). Molecular characteristics of diverse populations are consistent with the hypothesis of a recent invasion of *Drosophila melanogaster* by mobile *P* elements. *Mol. Biol. Evol.* **5,** 252–269.

Aravin, A. A., Hannon, G. J., and Brennecke, J. (2007). The Piwi-piRNA pathway provides an adaptive defense in the transposon arms race. *Science* **318,** 761–764.

Argumuganathan, K., and Earle, E. D. (1991). Nuclear DNA content of some important plant species. *Plant Mol. Biol. Rep.* **9,** 208–218.

Arnaud, L., and Haubruge, E. (2002). Insecticide resistance enhances male reproductive success in a beetle. *Evolution* **56,** 2435–2444.

Baker, J. E., Perezmendoza, J., Beeman, R. W., and Throne, J. E. (1998). Fitness of a malathion-resistant strain of the parasitoid *Anisopteromalus calandrae* (Hymenoptera: Pteromalidae). *J. Econ. Entomol.* **91,** 50–55.

Baker, M. B., Dastur, S. R., Jaffe, B. D., and Wong, T. (2008). Mating competition in Colorado potato beetles (Coleoptera: Chrysomelidae) does not show a cost of insecticide resistance. *Ann. Entomol. Soc. Am.* **101,** 371–377.

Bass, C., and Field, L. M. (2011). Gene amplification and insecticide resistance. *Pest Manag. Sci.* **67,** 886–890.

Bellen, H. J., Levis, R. W., He, Y. C., Carlson, J. W., Evans-Holm, M., Bae, E., Kim, J., Metaxakis, A., Savakis, C., Schulze, K. L., *et al.* (2011). The *Drosophila* gene disruption project: Progress using transposons with distinctive site-specificities. *Genetics* **188,** 731–743.

Bennetzen, J. (2005). Transposable elements, gene creation and genome rearrangement in flowering plants. *Curr. Opin. Genet. Dev.* **15,** 621–627.

Berrada, S., and Fournier, D. (1997). Transposition-mediated transcriptional overexpression as a mechanism of insecticide resistance. *Mol. Gen. Genet.* **256,** 348–354.

Berticat, C., Boquien, G., Raymond, M., and Chevillon, C. (2002). Insecticide resistance genes induce a mating competition cost in *Culex pipiens* mosquitoes. *Genet. Res.* **79,** 41–47.

Bielza, P., Quinto, V., Grávalos, C., Abellán, J., and Fernández, E. (2008). Lack of fitness costs of insecticide resistance in the western flower thrips (Thysanoptera: Thripidae). *J. Econ. Entomol.* **101,** 499–503.

Biémont, C. (2010). A brief history of the status of transposable elements: From junk DNA to major players in evolution. *Genetics* **186,** 1085–1093.

Biémont, C., and Vieira, C. (2006). Genetics—Junk DNA as an evolutionary force. *Nature* **443,** 521–524.

Biessmann, H., Valgeirsdottir, K., Lofsky, A., Chin, C., Ginther, B., Levis, R. W., and Pardue, M. L. (1992). HeTA, a transposable element specifically involved in "healing" broken chromosome ends in *Drosophila melanogaster*. *Mol. Cell. Biol.* **12,** 3910–3918.

Bloch, G., and Wool, D. (1994). Methidathion resistance in the sweet potato whitefly (Aleyrodidae: Homoptera) in Israel: Selection, heritability, and correlated changes of esterase activity. *J. Econ. Entomol.* **87,** 1147–1156.

Bogwitz, M. R., Chung, H., Magoc, L., Rigby, S., Wong, W., O'Keefe, M., McKenzie, J. A., Batterham, P., and Daborn, P. J. (2005). Cyp12a4 confers lufenuron resistance in a natural population of *Drosophila melanogaster*. *Proc. Natl. Acad. Sci. U.S.A.* **102,** 12807–12812.

Boivin, T., Chabert d'Hières, C., Bouvier, J. C., Beslay, D., and Sauphanor, B. (2001). Pleiotropy of insecticide resistance in the codling moth, *Cydia pomonella*. *Entomol. Exp. Appl.* **99,** 381–386.

Bonduriansky, R., and Chenoweth, S. F. (2009). Intralocus sexual conflict. *Trends Ecol. Evol.* **24,** 280–288.

Brewer, M. J., and Trumble, J. T. (1991). Inheritance and fitness consequences of resistance to fenvalerate in *Spodoptera exigua* (Lepidoptera, Noctuidae). *J. Econ. Entomol.* **84,** 1638–1644.

Brookfield, J. F. Y. (2005). The ecology of the genome—Mobile DNA elements and their hosts. *Nat. Rev. Genet.* **6,** 128–136.

Brookfield, J., and Badge, R. (1997). Population genetics models of transposable elements. *Genetica* **100,** 281–294.

Burt, A., and Trivers, R. (2006). Genes in Conflict: The Biology of Selfish Genetic Elements. The Belknap Press of Harvard University Press, London.

Carrière, Y., Deland, J.-P., Roff, D. A., and Vincent, C. (1994). Life-history costs associated with the evolution of insecticide resistance. *Proc. R. Soc. Lond. B Biol. Sci.* **258,** 35–40.

Carrière, Y., Roff, D. A., and Deland, J. P. (1995). The joint evolution of diapause and insecticide resistance—A test of an optimality model. *Ecology* **76,** 1497–1505.

Carrière, Y., Ellers-Kirk, C., Patin, A. L., Sims, M. A., Meyer, S., Liu, Y. B., Dennehy, T. J., and Tabashnik, B. E. (2001). Overwintering cost associated with resistance to transgenic cotton in the pink bollworm (Lepidoptera: Gelechiidae). *J. Econ. Entomol.* **94,** 935–941.

Castañeda, L. E., Barrientos, K., Cortes, P. A., Figueroa, C. C., Fuentes-Contreras, E., Luna-Rudloff, M., Silva, A. X., and Bacigalupe, L. D. (2011). Evaluating reproductive fitness and metabolic costs for insecticide resistance in *Myzus persicae* from Chile. *Physiol. Entomol.* **36,** 253–260.

Catania, F., Kauer, M. O., Daborn, P. J., Yen, J. L., ffrench-Constant, R. H., and Schlötterer, C. (2004). World-wide survey of an *Accord* insertion and its association with DDT resistance in *Drosophila melanogaster*. *Mol. Ecol.* **13,** 2491–2504.

Charlesworth, D., and Charlesworth, B. (1995). Transposable elements in inbreeding and outbreeding populations. *Genetics* **140,** 415–417.

Charlesworth, B., and Charlesworth, D. (2000). The degeneration of Y chromosomes. *Philos. Trans. R. Soc. Lond. B Biol. Sci.* **355,** 1563–1572.

Charlesworth, B., and Charlesworth, D. (2010). Elements of Evolutionary Genetics. Roberts and Company Publishers, Greenwood Village, Colorado.

Charlesworth, B., and Langley, C. H. (1986). The evolution of self-regulated transposition of transposable elements. *Genetics* **112,** 359–383.

Charlesworth, B., and Langley, C. H. (1989). The population genetics of *Drosophila* transposable elements. *Annu. Rev. Genet.* **23,** 251–287.

Charlesworth, B., Sniegowski, P., and Stephan, W. (1994). The evolutionary dynamics of repetitive DNA in eukaryotes. *Nature* **371,** 215–220.

Charlesworth, B., Langley, C., and Sniegowski, P. (1997). Transposable element distributions in *Drosophila*. *Genetics* **147,** 1993–1995.

Chen, S., and Li, X. C. (2007). Transposable elements are enriched within or in close proximity to xenobiotic-metabolizing cytochrome P450 genes. *BMC Evol. Biol.* **7,** 13.

Chevillon, C., Bourguet, D., Rousset, F., Pasteur, N., and Raymond, M. (1997). Pleiotropy of adaptive changes in populations: Comparisons among insecticide resistance genes in *Culex pipiens*. *Genet. Res.* **70,** 195–203.

Chung, H., Bogwitz, M. R., McCart, C., Andrianopoulos, A., ffrench-Constant, R. H., Batterham, P., and Daborn, P. J. (2007). Cis-regulatory elements in the *Accord* retrotransposon result in tissue-specific expression of the *Drosophila melanogaster* insecticide resistance gene *Cyp6g1*. *Genetics* **175,** 1071–1077.

Clark, J. M., and Yamaguchi, I. (eds.) (2002). *In* "Agrochemical Resistance: Extent, Mechanism, and Detection"*In* ACS Symposium Series No. 808. American Chemical Society, Washington, DC.

Cohan, F., King, E., and Zawadzki, P. (1994). Amelioration of the deleterious pleiotropic effects of an adaptive mutation in *Bacillus subtilis*. *Evolution* **48,** 81–95.

Cordaux, R., and Batzer, M. A. (2009). The impact of retrotransposons on human genome evolution. *Nat. Rev. Genet.* **10,** 691–703.

Crow, J. F. (1957). Genetics of insect resistance to chemicals. *Annu. Rev. Entomol.* **2,** 227–246.

Currie, D. B., Mackay, T. F., and Partridge, L. (1998). Pervasive effects of P element mutagenesis on body size in *Drosophila melanogaster*. *Genet. Res.* **72,** 19–24.

Daborn, P., Boundy, S., Yen, J., Pittendrigh, B., and ffrench-Constant, R. (2001). DDT resistance in *Drosophila* correlates with *Cyp6g1* over-expression and confers cross-resistance to the neonicotinoid imidacloprid. *Mol. Genet. Genomics* **266,** 556–563.

Daborn, P. J., Yen, J. L., Bogwitz, M. R., Le Goff, G., Feil, E. S., Jeffers, S., Tijet, N., Perry, T., Heckel, D., Batterham, P., Feyereisen, R., Wilson, T. G., *et al.* (2002). A single P450 allele associated with insecticide resistance in *Drosophila*. *Science* **297,** 2253–2256.

Daniels, S. B., Peterson, K. R., Strausbaugh, L. D., Kidwell, M. G., and Chovnick, A. (1990). Evidence for horizontal transmission of the *P*-transposable element between *Drosophila* species. *Genetics* **124,** 339–355.

Dapkus, D., and Merrell, D. J. (1977). Chromosomal analysis of DDT-resistance in a long-term selected population of *Drosophila melanogaster*. *Genetics* **87,** 685–697.

Delpuech, J.-M., Aquadro, C. F., and Roush, R. T. (1993). Noninvolvement of the long terminal repeat of transposable element *17.6* in insecticide in *Drosophila*. *Proc. Natl. Acad. Sci. U.S.A.* **90,** 5643–5647.

Denholm, I., Pickett, J. A., and Devonshire, A. L. (eds.) (1999). *In* "Insecticide Resistance: From Mechanisms to Management." CAB International Publishing, Wallingford, UK.

Devonshire, A. L., and Field, L. M. (1991). Gene amplification and insecticide resistance. *Annu. Rev. Entomol.* **36,** 1–23.

Dominguez, A., and Albornoz, J. (1996). Rates of movement of transposable elements in *Drosophila melanogaster*. *Mol. Gen. Genet.* **251,** 130–138.

Doolittle, W. F., and Sapienza, C. (1980). Selfish genes, the phenotype paradigm and genome evolution. *Nature* **284,** 601–603.

Dowsett, A., and Young, M. (1982). Differing levels of dispersed repetitive DNA among closely related species of *Drosophila*. *Proc. Natl. Acad. Sci. U.S.A.* **79,** 4570–4574.

Drnevich, J. M., Reedy, M. M., Ruedi, E. A., Rodriguez-Zas, S., and Hughes, K. A. (2004). Quantitative evolutionary genomics: Differential gene expression and male reproductive success in *Drosophila melanogaster*. *Proc. R. Soc. Lond. B Biol. Sci.* **271,** 2267–2273.

Duron, O., Labbé, P., Berticat, C., Rousset, F., Guillot, S., Raymond, M., and Weill, M. (2006). High *Wolbachia* density correlates with cost of infection for insecticide resistant *Culex pipiens* mosquitoes. *Evolution* **60,** 303–314.

Eanes, W. F., Wesley, C., Hey, J., Houle, D., and Ajioka, J. W. (1988). The fitness consequences of *P* element insertion in *Drosophila melanogaster*. *Genet. Res.* **52,** 17–26.

Emerson, J. J., Cardoso-Moreira, M., Borevitz, J. O., and Long, M. (2008). Natural selection shapes genome-wide patterns of copy-number polymorphism in *Drosophila melanogaster*. *Science* **320,** 1629–1631.

Feschotte, C. (2008). Transposable elements and the evolution of regulatory networks. *Nat. Rev. Genet.* **9,** 397–405.

Feyereisen, R. (1995). Molecular biology of insecticide resistance. *Toxicol. Lett.* **82,** 83–90.

ffrench-Constant, R. H. (1999). Target site mediated insecticide resistance: What questions remain? *Insect Biochem. Mol. Biol.* **29,** 397–403.

ffrench-Constant, R. H., Pittendrigh, B., Vaughan, A., and Anthony, N. (1998). Why are there so few resistance-associated mutations in insecticide target genes? *Philos. Trans. R. Soc. Lond. B Biol. Sci.* **353,** 1685–1693.

ffrench-Constant, R. H., Daborn, P. J., and Le Goff, G. (2004). The genetics and genomics of insecticide resistance. *Trends Genet.* **20,** 163–170.

ffrench-Constant, R., Daborn, P., and Feyereisen, R. (2006). Resistance and the jumping gene. *Bioessays* **28,** 6–8.

Field, L. M., Devonshire, A. L., and Forde, B. G. (1988). Molecular evidence that insecticide resistance in peach-potato aphids (*Myzus persicae* Sulz.) results from amplification of an esterase gene. *Biochem. J.* **251,** 309–312.

Finnegan, D. J. (1989). Eukaryotic transposable elements and genome evolution. *Trends Genet.* **5,** 103–107.

Finnegan, D. J. (1992). Transposable elements. *In* "The Genome of *Drosophila melanogaster*" (D. L. Lindsley and G. G. Zimm, eds.), pp. 1096–1107. Academic Press, San Diego.

Fisher, R. A. (1928). The possible modification of the responses of wild type to recurrent mutations. *Am. Nat.* **62,** 115–126.

Fitzpatrick, B., and Sved, J. (1986). High-levels of fitness modifiers induced by hybrid dysgenesis in *Drosophila melanogaster*. *Genet. Res.* **48,** 89–94.

Follett, P. A., Gould, F., and Kennedy, G. G. (1993). Comparative fitness of 3 strains of Colorado potato beetle (Coleoptera, Chrysomelidae) in the field—Spatial and temporal variation in insecticide selection. *J. Econ. Entomol.* **86,** 1324–1333.

Foster, S. P., Young, S., Williamson, M. S., Duce, I., Denholm, I., and Devine, G. J. (2003). Analogous pleiotropic effects of insecticide resistance genotypes in peach-potato aphids and houseflies. *Heredity* **91,** 98–106.

Gahan, L. J., Gould, F., and Heckel, D. G. (2001). Identification of a gene associated with Bt resistance in *Heliothis virescens*. *Science* **293,** 857–860.

George, J. A., Traverse, K. L., DeBaryshe, P. G., Kelley, K. J., and Pardue, M. L. (2010). Evolution of diverse mechanisms for protecting chromosome ends by *Drosophila* TART telomere retrotransposons. *Proc. Natl. Acad. Sci. U.S.A.* **107,** 21052–21057.

Georghiou, G. P., and Lagunes-Tejeda, A. (1991). The Occurrence of Resistance to Pesticides in Arthropods. An Index of Cases Reported Through 1989. Food and Agriculture Organization, Rome.

Georghiou, G. P., and Saito, T. (eds.) (1983). *In* "Pest Resistance to Pesticides" Plenum Press, New York.

González, J., and Petrov, D. A. (2009). The adaptive role of transposable elements in the *Drosophila* genome. *Gene* **448,** 124–133.

Grant, D. F., and Hammock, B. D. (1992). Genetic and molecular evidence for a trans-acting regulatory locus controlling glutathione stransferase-2 expression in *Aedes aegypti*. *Mol. Gen. Genet.* **234,** 169–176.

Green, M. (1988). Mobile DNA elements and spontaneous gene mutation. *In* "Eukaryotic Transposable Elements as Mutagenic Agents" (M. Lambert, J. McDonald, and I. Weinstein, eds.), pp. 41–50. Cold Spring Harbor Laboratory Press, Cold Spring Harbor.

Haldane, J. B. S. (1932). The Causes of Evolution. Harper, New York, NY.

Hallstrom, I. (1985). Genetic regulation of the cytochrome P-450 system in *Drosophila melanogaster*. II. Localization of some genes regulating cytochrome P-450 activity. *Chem. Biol. Interact.* **56,** 173–184.

Harano, T., Okada, K., Nakayama, S., Miyatake, T., and Hosken, D. J. (2010). Intralocus sexual conflict unresolved by sex-limited trait expression. *Curr. Biol.* **20,** 2036–2039.

Hardstone, M. C., Komagata, O., Kasai, S., Tomita, T., and Scott, J. G. (2010). Use of isogenic strains indicates CYP9M10 is linked to permethrin resistance in *Culex pipiens quinquefasciatus*. *Insect Mol. Biol.* **19,** 717–726.

Hartl, D. L., and Clark, A. G. (1997). Principles of Population Genetics. Sinauer Associates, Sunderland, Massachusetts.

Haubruge, E., and Arnaud, L. (2001). Fitness consequences of malathion-specific resistance in red flour beetle (Coleoptera: Tenebrionidae) and selection for resistance in the absence of malathion. *J. Econ. Entomol.* **94,** 552–557.

Hickey, D. A. (1982). Selfish DNA: A sexually-transmitted nuclear parasite. *Genetics* **101,** 519–531.

Hilgenboecker, K., Hammerstein, P., Schlattmann, P., Telschow, A., and Werren, J. H. (2008). How many species are infected with *Wolbachia*? A statistical analysis of current data. *FEMS Microbiol. Lett.* **281,** 215–220.

Hollingsworth, R. G., Tabashnik, B. E., Johnson, M. W., Messing, R. H., and Ullman, D. E. (1997). Relationship between susceptibility to insecticides and fecundity across populations of cotton aphid (Homoptera: Aphididae). *J. Econ. Entomol.* **90,** 55–58.

Hollingworth, R. M., and Dong, K. (2008). The biochemical and molecular genetic basis of resistance to pesticides in arthropods. *In* "Global Pesticide Resistance in Arthropods" (M. E. Whalon, D. Mota-Sanchez, and R. M. Hollingworth, eds.), pp. 40–89. CABI, Oxfordshire, UK.

Hosken, D. J., Stockley, P., Tregenza, T., and Wedell, N. (2009). Monogamy and the battle of the sexes. *Annu. Rev. Entomol.* **54,** 361–378.

Hua-Van, A., Le Rouzic, A., Boutin, T. S., Filée, J., and Capy, P. (2011). The struggle for life of the genome's selfish architects. *Biol. Direct* **6,** 19.

Hurst, G. D. D., and Werren, J. H. (2001). The role of selfish genetic elements in eukaryotic evolution. *Nat. Rev. Genet.* **2,** 597–606.

Ishaaya, I. (ed.) (2001). *In* "Biochemical Sites of Insecticide Action and Resistance" Springer, New York.

Itokawa, K., Komagata, O., Kasai, S., Okamura, Y., Masada, M., and Tomita, T. (2010). Genomic structures of *Cyp9m10* in pyrethroid resistant and susceptible strains of *Culex quinquefasciatus*. *Insect Biochem. Mol. Biol.* **40,** 631–640.

Itokawa, K., Komagata, O., Kasai, S., Masada, M., and Tomita, T. (2011). Cis-acting mutation and duplication: History of molecular evolution in a P450 haplotype responsible for insecticide resistance in *Culex quinquefasciatus*. *Insect Biochem. Mol. Biol.* **41,** 503–512.

Janmaat, A. F., and Myers, J. H. (2005). The cost of resistance to *Bacillus thuringiensis* varies with the host plant of *Trichoplusia ni*. *Proc. R. Soc. Lond. B Biol. Sci.* **272,** 1031–1038.

Jiang, N., Bao, Z. R., Zhang, X. Y., Eddy, S. R., and Wessler, S. R. (2004). Pack–MULE transposable elements mediate gene evolution in plants. *Nature* **431,** 569–573.

Kaminker, J. S., Bergman, C. M., Kronmiller, B., Carlson, J., Svirskas, R., Patel, S., Frise, E., Wheeler, D. A., Lewis, S. E., Rubin, G. M., Ashburner, M., and Celniker, S. E. (2002). The transposable elements of the *Drosophila melanogaster* euchromatin: A genomics perspective. *Genome Biol.*3**RESEARCH0084.**

Karunaratne, S. H., Jayawardena, K. G., Hemingway, J., and Ketterman, A. J. (1993). Characterization of a B-type esterase involved in insecticide resistance from the mosquito *Culex quinquefasciatus*. *Biochem. J.* **294,** 575–579.

Kidwell, M., and Lisch, D. (2001). Perspective: Transposable elements, parasitic DNA, and genome evolution. *Evolution* **55,** 1–24.

Kikkawa, H. (1961). Genetical studies on the resistance to parathion in *Drosophila melanogaster*. *Annu. Rep. Sci. Works Fac. Sci. Osaka Univ.* **9,** 1–20.

Kim, J., Vanguri, S., Boeke, J. D., Gabriel, A., and Voytas, D. F. (1998). Transposable elements and genome organization: A comprehensive survey of retrotransposons revealed by the complete *Saccharomyces cerevisiae* genome sequence. *Genome Res.* **8,** 464–478.

Labbé, P., Sidos, N., Raymond, M., and Lenormand, T. (2009). Resistance gene replacement in the mosquito *Culex pipiens*: Fitness estimation from long-term cline series. *Genetics* **182,** 303–312.

Lande, R. (1983). The response to selection on major and minor mutations affecting a metrical trait. *Heredity* **50,** 47–65.

Lander, E. S., Linton, L. M., Birren, B., Nusbaum, C., Zody, M. C., Baldwin, J., Devon, K., Dewar, K., Doyle, M., FitzHugh, W., *et al.* (2001). Initial sequencing and analysis of the human genome. *Nature* **409,** 860–921.

Levin, H. L., and Moran, J. V. (2011). Dynamic interactions between transposable elements and their hosts. *Nat. Rev. Genet.* **12,** 615–627.

Levis, R. W., Ganesan, R., Houtchens, K., Tolar, L. A., and Sheen, F. M. (1993). Transposons in place of telomeric repeats at a *Drosophila* telomere. *Cell* **75,** 1083–1093.

Li, X., Schuler, M. A., and Berenbaum, M. R. (2007). Molecular mechanisms of metabolic resistance to synthetic and natural xenobiotics. *Annu. Rev. Entomol.* **52,** 231–253.

Lockton, S., and Gaut, B. S. (2010). The evolution of transposable elements in natural populations of self-fertilizing *Arabidopsis thaliana* and its outcrossing relative *Arabidopsis lyrata*. *BMC Evol. Biol.* **10,** 10.

Lockton, S., Ross-Ibarra, J., and Gaut, B. S. (2008). Demography and weak selection drive patterns of transposable element diversity in natural populations of *Arabidopsis lyrata*. *Proc. Natl. Acad. Sci. U.S.A.* **105,** 13965–13970.

Lynch, M., and Conery, J. (2003). The origins of genome complexity. *Science* **302,** 1401–1404.

Mackay, T. F. (1986). Transposable element-induced fitness mutations in *Drosophila melanogaster*. *Genet. Res.* **48,** 77–87.

Mackay, T. F. (1989). Transposable elements and fitness in *Drosophila melanogaster*. *Genome* **31,** 284–295.

Mackay, T. F., Lyman, R. F., and Jackson, M. S. (1992). Effects of *P* element insertions on quantitative traits in *Drosophila melanogaster*. *Genetics* **130,** 315–332.

Macnair, M. R. (1991). Why the evolution of resistance to anthropogenic toxins normally involves major gene changes: The limits to natural selection. *Genetica* **84,** 213–219.

Mallet, J. (1989). The evolution of insecticide resistance: Have the insects won? *Trends Ecol. Evol.* **4,** 336–340.

Malone, C. D., and Hannon, G. J. (2009). Small RNAs as guardians of the genome. *Cell* **136,** 656–668.

Maside, X., Assimacopoulos, S., and Charlesworth, B. (2000). Rates of movement of transposable elements on the second chromosome of *Drosophila melanogaster*. *Genet. Res.* **75,** 275–284.

Maside, X., Bartolomé, C., and Charlesworth, B. (2002). *S*-element insertions are associated with the evolution of the Hsp70 genes in *Drosophila melanogaster*. *Curr. Biol.* **12,** 1686–1691.

Mason, P. L. (1998). Selection for and against resistance to insecticides in the absence of insecticide: A case study of malathion resistance in the saw-toothed grain beetle, *Oryzaephilus surinamensis* (Coleoptera: Silvanidae). *Bull. Entomol. Res.* **88,** 177–188.

May, R. (1985). Population biology—Evolution of pesticide resistance. *Nature* **315,** 12–13.

McCart, C., Buckling, A., and ffrench-Constant, R. H. (2005). DDT resistance in flies carries no cost. *Curr. Biol.* **15,** R587–R589.

McClintock, B. (1950). The origin and behavior of mutable loci in maize. *Proc. Natl. Acad. Sci. U.S.A.* **36,** 344–355.

McClintock, B. (1984). The significance of responses of the genome to challenge. *Science* **226,** 792–801.

McKenzie, J. A., and Batterham, P. (1994). The genetics, molecular and phenotypic consequences of selection for insecticide resistance. *Trends Ecol. Evol.* **9,** 166–169.

Melander, A. L. (1914). Can insects become resistant to sprays? *J. Econ. Entomol.* **7,** 167–173.

Messing, J., and Dooner, H. (2006). Organization and variability of the maize genome. *Curr. Opin. Plant Biol.* **9,** 157–163.

Minkoff, C., and Wilson, T. G. (1992). The competitive ability and fitness components of the Methoprene-tolerant (Met) *Drosophila* mutant resistant to juvenile hormone analog insecticides. *Genetics* **131,** 91–97.

Mouchès, C., Pasteur, N., Berge, J. B., Hyrien, O., Raymond, M., de Saint Vincent, B. R., de Silvestri, M., and Georghiou, G. P. (1986). Amplification of an esterase gene is responsible for insecticide resistance in a California *Culex* mosquito. *Science* **233,** 778–780.

Mouchès, C., Pauplin, Y., Agarwal, M., Lemieux, L., Herzog, M., Abadon, M., Beyssat-Arnaouty, V., Hyrien, O., de Saint Vincent, B. R., Georghiou, G. P., and Pasteur, N. (1990). Characterization of amplification core and Esterase-b1 gene responsible for insecticide resistance in *Culex*. *Proc. Natl. Acad. Sci. U.S.A.* **87,** 2574–2578.

Nuzhdin, S., and Mackay, T. (1995). The genomic rate of transposable element movement in *Drosophila melanogaster*. *Mol. Biol. Evol.* **12,** 180–181.

O'Donnell, K. A., and Burns, K. H. (2010). Mobilizing diversity: Transposable element insertions in genetic variation and disease. *Mob. DNA* **1**(1), 21.

Oakeshott, J. G., Horne, I., Sutherland, T. D., and Russell, R. J. (2003). The genomics of insecticide resistance. *Genome Biol.* **4,** 202.

Omer, A. D., Leigh, T. F., and Granett, J. (1992). Insecticide resistance of greenhouse-whitefly (Homoptera, Aleyrodidae) and fitness on plant hosts relative to the San-Joaquin valley (California) cotton agroecosystem. *J. Appl. Entomol.* **113,** 244–251.

Oppert, B., Hammel, R., Throne, J. E., and Kramer, K. J. (2000). Fitness costs of resistance to *Bacillus thuringiensis* in the Indian meal moth, *Plodia interpunctella*. *Entomol. Exp. Appl.* **96,** 281–287.

Orgel, L. E., and Crick, F. H. (1980). Selfish DNA: The ultimate parasite. *Nature* **284,** 604–607.

Petrov, D. A., Fiston-Lavier, A.-S., Lipatov, M., Lenkov, K., and González, J. (2011). Population genomics of transposable elements in *Drosophila melanogaster*. *Mol. Biol. Evol.* **28,** 1633–1644.

Raymond, M., Beyssat-Arnaouty, V., Sivasubramanian, N., Mouchès, C., Georghiou, G. P., and Pasteur, N. (1989). Amplification of various esterase B's responsible for organophosphate resistance in *Culex* mosquitoes. *Biochem. Genet.* **27,** 417–423.

Raymond, B., Sayyed, A. H., and Wright, D. J. (2005). Genes and environment interact to determine the fitness costs of resistance to *Bacillus thuringiensis*. *Proc. R. Soc. Lond. B Biol. Sci.* **272,** 1519–1524.

Raymond, B., Sayyed, A. H., Hails, R. S., and Wright, D. J. (2007). Exploiting pathogens and their impact on fitness costs to manage the evolution of resistance to *Bacillus thuringiensis*. *J. Appl. Ecol.* **44,** 768–780.

Raymond, B., Wright, D. J., and Bonsall, M. B. (2011). Effects of host plant and genetic background on the fitness costs of resistance to *Bacillus thuringiensis*. *Heredity* **106,** 281–288.

Rivero, A., Magaud, A., Nicot, A., and Vezilier, J. (2011). Energetic cost of insecticide resistance in *Culex pipiens* mosquitoes. *J. Med. Entomol.* **48,** 694–700.

Robertson, H. M., and Engels, W. R. (1989). Modified P elements that mimic the P cytotype in *Drosophila melanogaster*. *Genetics* **123,** 815–824.

Roush, R. T., and McKenzie, J. A. (1987). Ecological genetics of insecticide and acaricide resistance. *Annu. Rev. Entomol.* **32,** 361–380.

Schlenke, T. A., and Begun, D. J. (2004). Strong selective sweep associated with a transposon insertion in *Drosophila simulans*. *Proc. Natl. Acad. Sci. U.S.A.* **101,** 1626–1631.

Schmidt, J. M., Good, R. T., Appleton, B., Sherrard, J., Raymant, G. C., Bogwitz, M. R., Martin, J., Daborn, P. J., Goddard, M. E., Batterham, P., *et al.* (2010). Copy number variation and transposable elements feature in recent, ongoing adaptation at the *Cyp6g1* locus. *PLoS Genet.* **6,** e1000998.

Shapiro, J. A. (1969). Mutations caused by the insertion of genetic material into the galactose operon of *Escherichia coli*. *J. Mol. Biol.* **40,** 93–105.

Smith, D. T., Hosken, D. J., Rostant, W. G., Yeo, M., Griffin, R. M., Bretman, A., Price, T. A. R., ffrench-Constant, R. H., and Wedell, N. (2011). DDT resistance, epistasis and male fitness in flies. *J. Evol. Biol.* **24,** 1351–1362.

Sparks, T. C., Lockwood, J. A., Byford, R. L., Graves, J. B., and Leonard, B. R. (1989). The role of behavior in insecticide resistance. *Pestic. Sci.* **26,** 383–399.

Steinemann, M., and Steinemann, S. (1998). Enigma of Y chromosome degeneration: Neo-Y and Neo-X chromosomes of *Drosophila miranda* a model for sex chromosome evolution. *Genetica* **102,** 409–420.

Stouthamer, R., Breeuwer, J., and Hurst, G. (1999). *Wolbachia pipientis*: Microbial manipulator of arthropod reproduction. *Annu. Rev. Microbiol.* **53,** 71–102.

Tang, J. D., Gilboa, S., Roush, R. T., and Shelton, A. M. (1997). Inheritance, stability, lack of fitness costs of field-selected resistance to *Bacillus thuringiensis* in diamondback moth (Lepidoptera: Plutellidae) from Florida. *J. Econ. Entomol.* **90,** 732–741.

Tang, J. D., Collins, H. L., Roush, R. T., Metz, T. D., Earle, E. D., and Shelton, A. M. (1999). Survival, weight gain, and oviposition of resistant and susceptible *Plutella xylostella* (Lepidoptera: Plutellidae) on broccoli expressing CryIAc toxin of *Bacillus thuringiensis*. *J. Econ. Entomol.* **92,** 47–55.

Thomas, C. (1971). Genetic organization of chromosomes. *Annu. Rev. Genet.* **5,** 237–256.

Thompson, M., Steichen, J. C., and ffrench-Constant, R. H. (1993). Conservation of cyclodiene insecticide resistance associated mutations in insects. *Insect Mol. Biol.* **2,** 149–154.

Touchon, M., and Rocha, E. P. C. (2007). Causes of insertion sequences abundance in prokaryotic genomes. *Mol. Biol. Evol.* **24,** 969–981.

van Rij, R. P., and Berezikov, E. (2009). Small RNAs and the control of transposons and viruses in *Drosophila*. *Trends Microbiol.* **17,** 163–171.

Waters, L. C., Zelhof, A. C., Shaw, B. J., and Ch'ang, L.-Y. (1992). Possible involvement of the long terminal repeat of transposable element *17.6* in regulating expression of an insecticide resistance-associated P450 gene in *Drosophila*. *Proc. Natl. Acad. Sci. U.S.A.* **89,** 4855–4859.

Werren, J. H. (1997). Biology of *Wolbachia*. *Annu. Rev. Entomol.* **42,** 587–609.

Werren, J. H. (2011). Selfish genetic elements, genetic conflict, and evolutionary innovation. *Proc. Natl. Acad. Sci. U.S.A.* **108,** 10863–10870.

Whalon, M. E., Mota-Sanchez, D., and Hollingworth, R. M. (eds.) (2008). *In* "Global Pesticide Resistance in Arthropods." CAB International Publishing, Wallingford, UK.

White, N. D. G., and Bell, R. J. (1995). A malathion resistance gene associated with increased life-span of the rusty grain beetle, *Cryptolestes ferrugineus* (Coleoptera, Cucujidae). *J. Gerontol. A Biol. Sci. Med. Sci.* **50,** B9–B13.

Wicker, T., Sabot, F., Hua-Van, A., Bennetzen, J. L., Capy, P., Chalhoub, B., Flavell, A., Leroy, P., Morgante, M., Panaud, O., *et al.* (2007). A unified classification system for eukaryotic transposable elements. *Nat. Rev. Genet.* **8,** 973–982.

Wilson, T. G. (1993). Transposable elements as initiators of insecticide resistance. *J. Econ. Entomol.* **86,** 645–651.

Wilson, T. G. (2001). Resistance of *Drosophila* to toxins. *Annu. Rev. Entomol.* **46,** 545–571.

Wilson, T., and Turner, C. (1992). Molecular analysis of methoprene-tolerant, a gene in *Drosophila* involved in resistance to juvenile-hormone analog insect growth-regulators. *In* "Molecular Mechanisms of Insecticide Resistance: Diversity Among Insects" (C. A. Mullin and J. G. Scott, eds.), ACS Symposium Series No. 505, pp. 99–112. American Chemical Society, Washington, DC.

Xiao, H., Jiang, N., Schaffner, E., Stockinger, E. J., and van der Knaap, E. (2008). A retrotransposon-mediated gene duplication underlies morphological variation of tomato fruit. *Science* **319,** 1527–1530.

Xing, J. C., Zhang, Y. H., Han, K., Salem, A. H., Sen, S. K., Huff, C. D., Zhou, Q., Kirkness, E. F., Levy, S., Batzer, M. A., *et al.* (2009). Mobile elements create structural variation: Analysis of a complete human genome. *Genome Res.* **19,** 1516–1526.

Yamamoto, A., Yoneda, H., Hatano, R., and Asada, M. (1995). Influence of hexythiazox resistance on life-history parameters in the citrus red mite, *Panonychus citri* (McGregor). *J. Pestic. Sci.* **20,** 521–527.

Yang, S., Arguello, J. R., Li, X., Ding, Y., Zhou, Q., Chen, Y., Zhang, Y., Zhao, R., Brunet, F., Peng, L., *et al.* (2008). Repetitive element-mediated recombination as a mechanism for new gene origination in *Drosophila*. *PLoS Genet.* **4,** e3.

Zhang, Z., and Saier, M. H. (2009). A novel mechanism of transposon-mediated gene activation. *PLoS Genet.* **5,** e1000689.

Index

Note: Page numbers followed by "*f*" indicate figures.